IB Maths AI

Internal Assessment

IB Maths AI

Internal Assessment

The Definitive Application and Interpretation IA
Guide For the International Baccalaureate Diploma

Mudassir Mehmood

Zouev Elite IB Publishing

Published 2022

Printed by Zouev Elite IB Publishing

ISBN 978-1-9996115-6-9, paperback.

TABLE OF CONTENTS

PART I
THE MATH AI IA GUIDE

1. THE COMPLETE GUIDE FOR THE PERFECT MATH IA

The IA is a short-written report in which the students pick a certain topic of their interest and build on their mathematical concepts. There is no specific difference in an A&A and A&I other than criterion E which mainly lies in the domain of SL and HL respectively. IA is worth 20% of the entire grade. The supervisor is allowed to provide feedback once on the first draft. The final draft is marked by the supervisor and moderated by the IB. IA is marked on a total of 5 criterions.

Assessment Criteria:

A Presentation   4

B Mathematical communication   4

C Personal engagement   3

D Reflection   3

E Use of mathematics   6

How does IA differ from other math Assessments?

Other math Assessment

- Student has no option to choose a question
- There is one correct answer
- There is a one or few acceptable methods
- There is no need to explain why you are doing what you are doing.
- Limited amount of time
- Questions will be asked from any topic

Math IA

- Students choose the topic you want to answer.
- There is no single answer to the problem
- Students must justify everything what they do in the IA, as marks are awarded based on student's justification.
- Unlimited time (Approx. 30 hours)
- Should focus on the topic of student's interest.

A MUST TO DO EXERCISE BEFORE STARTING THE IA

Prior to start working on your IA, make sure you have a good understanding of the IA components, the assessment criterions and you should have read at least 3 IAs (low scoring, average scoring, and a high scoring). You should evaluate them against the assessment criterions and compare their actual score with the one you awarded against each criterion. Try to figure out where and why is the difference between the two scores.

Selection of the Topic

Selection of a topic is the trickiest process when it comes to AA SL/HL or AI HL math. One should spend good time on it and should discuss it with the teacher and peers. Selection of topic should be divided into these steps.

Step 1 – Mind Mapping

Thinking of what stimulate you, it could be something you have been thinking through your childhood, or something you recently have come across. It could be something that interests you, could be your hobby or idea you heard from someone else. Make list of your key words.

Here is an example but the list can go on and on....

Physics	Communication	Radiations
Chemistry	Internet	Rumor
Biology	Games	Computers
Geography	Forest	Algorithms
Cell phone	Tablet	Logs
Signals	Cricket	Water flow
Burgers	Football	Sound
Fruits	Lottery	Art
Light	Euler	Symmetry
Patterns	Reiman's sum	Likelihood
Correlation	IT	Geometry
Medal	Millionaire	Correlations
Buildings	Virus	Bridges
Structure	Bacteria	Paintings
Movies	Radioactivity	Orbits

Start mind mapping using these key words with each key word being the source of generating more thoughts and idea that can be linked together later. Do not focus on just one point instead think freely and develop connections as much as you can.

Start connection these ideas with the five topic/s in math i.e., numbers and patterns, functions, geometry and trigonometry, statistics and probability and calculus. Your key words may come under so many topics in math, do not worry, it is good.

List of Topics.

Whenever it comes to take ideas for picking up a topi/question/problem for the IA, I use IB Math resources. Here is the link (https://ibmathsresources.com/maths-ia-maths-exploration-topics/). You can pick any of the topics given below and start your working. I am writing few questions here. There are more than 200 topics/questions available on this website. You can visit this link to access these.

Algebra and number

1) Modular arithmetic - This technique is used throughout Number Theory. For example, Mod 3 means the remainder when dividing by 3.

2) Goldbach's conjecture: "Every even number greater than 2 can be expressed as the sum of two primes." One of the great unsolved problems in mathematics.

3) Probabilistic number theory

4) Applications of complex numbers: The stunning graphics of Mandelbrot and Julia Sets are generated by complex numbers.

Geometry

1a) Non-Euclidean geometries: This allows us to "break" the rules of conventional geometry – for example, angles in a triangle no longer add up to 180 degrees. In some geometries triangles add up to more than 180 degrees, in others less than 180 degrees.

1b) The shape of the universe – non-Euclidean Geometry is at the heart of Einstein's theories on General Relativity and essential to understanding the shape and behaviour of the universe.

2) Hexaflexagons: These are origami style shapes that through folding can reveal extra faces.

3) Minimal surfaces and soap bubbles: Soap bubbles assume the minimum possible surface area to contain a given volume.

4) Tesseract – a 4D cube: How we can use maths to imagine higher dimensions.

Statistics and modelling 1 [topics could be studied in-depth]

1) Traffic flow: How maths can model traffic on the roads.

2) Logistic function and constrained growth

3) Benford's Law – using statistics to catch criminals by making use of a surprising distribution.

Statistics and modelling 2 [more simplistic topics: correlation, normal, Chi squared]

1) grades? Studies have shown that a good night's sleep raises academic attainment.

2) Is there a correlation between height and weight? (pdf). The NHS use a chart to decide what someone should weigh depending on their height. Does this mean that height is a good indicator of weight?

3) Is there a correlation between arm span and foot height? This is also a potential opportunity to discuss the Golden Ratio in nature.

4) Is there a correlation between smoking and lung capacity?

5) Is there a correlation between GDP and life expectancy? Run the Gap minder graph to show the changing relationship between GDP and life expectancy over the past few decades.

Games and game theory

1) The prisoner's dilemma: The use of game theory in psychology and economics.

2) Sudoku

3) Gambler's fallacy: A good chance to investigate misconceptions in probability and probabilities in gambling. Why does the house always win?

4) Bluffing in Poker: How probability and game theory can be used to explore the the best strategies for bluffing in poker.

Topology and networks

1) Knots

2) Steiner problem

3) Chinese postman problem – This is a problem from graph theory – how can a postman deliver letters to every house on his streets in the shortest time possible?

4) Travelling salesman problem

Mathematics and Physics

1) The Monkey and the Hunter – How to Shoot a Monkey – Using Newtonian mathematics to decide where to aim when shooting a monkey in a tree.

2) How to Design a Parachute – looking at the physics behind parachute design to ensure a safe landing!

3) Galileo: Throwing cannonballs off The Leaning Tower of Pisa – Recreating Galileo's classic experiment, and using maths to understand the surprising result.

Maths and computing

1) The Van Eck Sequence – The Van Eck Sequence is a sequence that we still don't fully understand – we can use programming to help!

2) <u>Solving maths problems using computers</u> – computers are useful in solving mathematical problems. Here are some examples solved using Python.

3) <u>Stacking cannonballs – solving maths with code</u> – how to stack cannonballs in different configurations.

Further ideas:

1) <u>Radiocarbon dating</u> – understanding radioactive decay allows scientists and historians to accurately work out something's age – whether it be from thousands or even millions of years ago.

2) <u>Gravity, orbits, and escape velocity</u> – Escape velocity is the speed required to break free from a body's gravitational pull. Essential knowledge for future astronauts.

3) <u>Mathematical methods in economics</u> – maths is essential in both business and economics – explore some economics-based maths problems.

Mind Map

You can start your IA even in year 1, however, the best time to start thinking about your IA is when you are done with at least 80% of your math syllabus as you will have enough of the options available to pick a topic from. All the math topics should be on your tips so that you exactly know what stimulate you should be handled using a particular math topic. Here is the academic mind map for both AI SL/HL and AA SL/HL. Before finalizing your topic of interest, go through these mind maps and make sure you have a solution to your question/problem available in one/few of these topics.

Number and Algebra

ERROR AND APPROXIMATION

$\%\text{ age erro} = \left| \dfrac{V_A - V_E}{V_E} \right| \times 100\%$

V_E = Exact Value

V_A = Approximate Value

Significant figures

0.002 (1 s.f)

1.002 (4 s.f)

4.0000 (5 s.f)

Decimal places

23.2 (1 d.p)

4.50 (2d.p)

compound interest $FV = PV\left(1 + \dfrac{r\%}{k}\right)^{nk}$

±FV = Future value

FINANCE

±PV = Present value

$r\%$ = interest (-ive for depreciation)

$k = C/Y$ $\left(\begin{array}{c} k=1 \\ \text{yearly,} \end{array}\begin{array}{c} k=4 \\ \text{quarterly,} \end{array}\begin{array}{c} k=12 \\ \text{monthly,} \end{array}\right.$
     $\left.\begin{array}{c} \\ \text{weekly,} \end{array}\begin{array}{c} \\ \text{daily} \end{array}\right).$
    $k=52$ $k=365$

P/Y   Payment per year   PMT → Payment

Amortization

Ben borrowed a loan of $130,000 from bank which would be returned in 25 years with an interest rate at 7% compounded monthly. What will be the monthly installment?

$N = 12 \times 25 = 300$ months

$I = 7\%$

$PV = 130,000$

$FV = 0$

$PMT = ?$ $\$ -919$

$P/Y = 12$

$C/Y = 12$

Functions input → process → output x^2 $f(x)=y$

LINEAR FUNCTION
$y = ax + b$
x - axis
$ax + by + c = 0$
$y = mx + c$
$y - y_1 = m(x - x_1)$

$f(x) = ax^2 + bx + c$ (standard form)

$a > 0$ V
$a < 0$ ∧

QUADRATICS

$f(x) = a(x-h)^2 + k$
VERTEX FORM
(h, k)

y-intercept
$x = 0$

$(P, 0)$ $(0, C)$ $(q, 0)$

x-intercepts
$y = 0$

$f(x) = a(x-p)(x-q)$
FACTOR FORM

DISCRIMINANT
$\Delta = b^2 - 4ac$

$\Delta < 0$
no real root

$\Delta = 0$
Repeated root

$\Delta > 0$
Two distinct Real roots

axis of symmetry
$x = -b/2a$

$y = a^x$
Inverses
$\log_a x$

$y = 2$ (0,0.67) x = 3/2 (0.5,0)

RATIONAL FUNCTIONS
$f(x) = \dfrac{4x-2}{2x-3}$

Horizontal asymptote
as $x \to 0$ then $y = 2$

vertical asymptote
$2x - 3 = 0$
$2x = 3$
$x = 3/2$

EXPONENTIAL AND LOGARITHMIC FUNCTIONS

$y = a^x$ $y = x$

$y = \log_a x$

COMPOSITE FUNCTION
$x \to g(x) \to f(g(x))$
   g   f

$f \circ g \circ x = f(g(x))$

2 → 4
3 → 9   one to one

2 → 4
-2 → 4   Many to one

$f \circ f^{-1}(x) = f^{-1} \circ f(x) = x$

$a f(x)$
a

TRANSFORMATION
$-f(x)$

$f(x-h) + k$

Geometry and Trigonometry

angle of elevation

angle of depression

M (6,10)
A (2,8)

$m = \dfrac{10-8}{6-2} = \dfrac{2}{4} = \dfrac{1}{2}$

$M = \left(\dfrac{6+2}{2}, \dfrac{10+8}{2}\right)$
$= (4, 9)$

Gradient of L (perp bisector)
$= -\dfrac{1}{\frac{1}{2}} = -2$

slope-intercept form
$y = mx + c$
$y = \dfrac{1}{2}x + c$
$10 = \dfrac{1}{2} \times 6 + c$
$10 = 3 + c$
$c = 7$
$y = \dfrac{1}{2}x + 7$

Toxic waste problem

Minor arc

arc length
$l = \dfrac{\theta}{360} \times 2\pi r$

major arc

Area of sector $= \dfrac{\theta}{360} \times \pi r^2$

VORONOI DIAGRAM

Edges
sites
vertices
cells

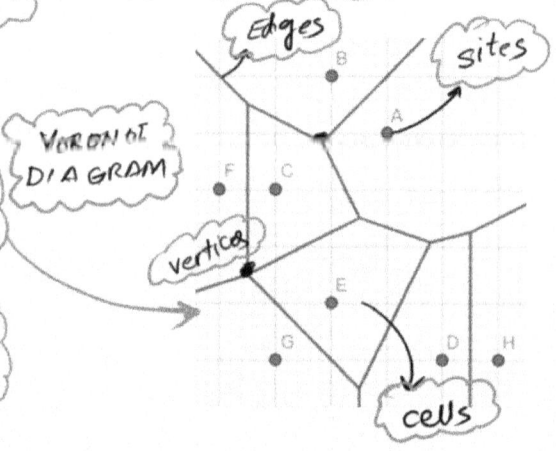

Statistics and Probability

Weekly Study Hours	Math Score %
8	62
12	75
5	50
15	80
16	75
18	88
5	80

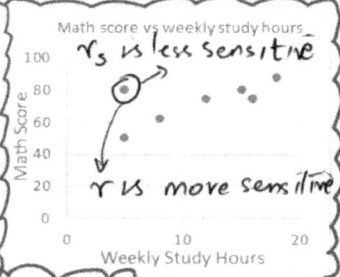

Math score vs weekly study hours

r_s is less sensitive

r is more sensitive

(scatter plot: Math Score vs Weekly Study Hours)

Rank Hours	Rank Score
5	6
4	4.5
6.5	7
3	2.5
2	4.5
1	1
6.5	2.5

$$PPMMC = r = 0.669$$

$$SMRC = r_s = 0.599$$

For equal values
$$\frac{6^{th} + 7^{th}}{2} = 6.5$$

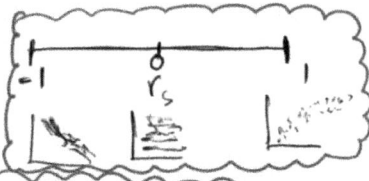

r_s scale from -1 to 1

$H_0 =$ Null Hyp
$H_a =$ Alternate Hyp
Significance level
e.g 5%

HYPOTHESIS TESTING

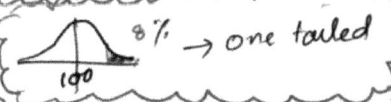

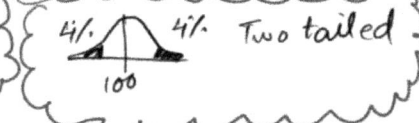

$8\% \rightarrow$ one tailed

4% ... 4% Two tailed

χ^2-test GOF
$H_0 : \mu_1 = \mu_2$
$Ha: \mu_1 \neq \mu_2$ or $\mu_1 > \mu_2$
or $\mu_1 < \mu_2$
$P > \alpha$ accept H_0
$P < \alpha$ reject H_0

χ^2-test
for independence
$d.f = ($no of rows$-1)$
$\times$
$($no of column$-1)$
$\chi^2 > C.V$
Reject
H_0

Calculus

$$\frac{d}{dx}(C) = 0$$
$$\frac{d}{dx}(cx) = C$$
$$\frac{d}{dx}(x) = 1$$
$$\frac{d}{dx}x^n = nx^{n-1}$$

e.g $y = 4x^3 + x^2 + x - 7$
$$\frac{dy}{dx} = \frac{d}{dx}(4x^3) + \frac{d}{dx}x^2 + \frac{d}{dx}x - \frac{d}{dx}(7)$$
$$= 4(3x^2) + 2x + 1 - 0$$
$$= 12x^2 + 2x + 1$$

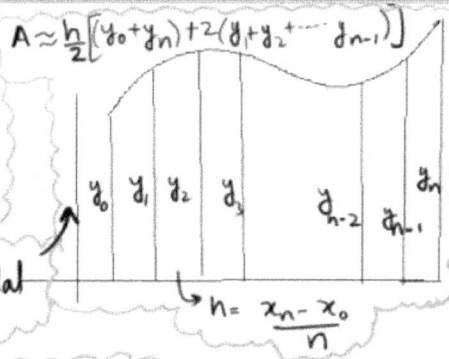

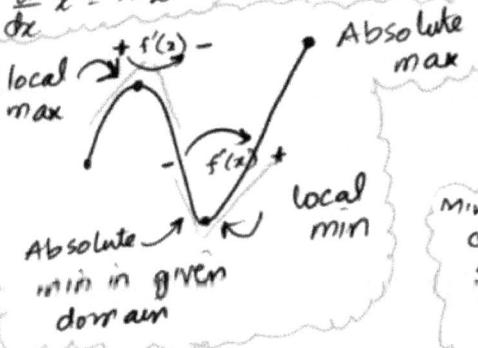

$$A \approx \frac{h}{2}\left[(y_0 + y_n) + 2(y_1 + y_2 + \cdots y_{n-1})\right]$$

Trapezoidal Rule

$$h = \frac{x_n - x_0}{n}$$

local max, Absolute max, $+f'(x)-$, local min, Absolute min in given domain

Minimizing cost S.A Time

where $\frac{dy}{dx} = 0$

Optimization

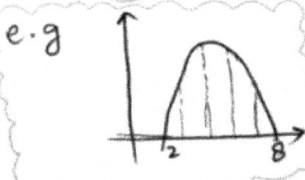

Maximizing profit Volume Sales

e.g (graph from 2 to 8)

$$h = \frac{8-2}{5} = 1.2$$

$$\int_2^8 y\,dx \approx \frac{1}{2}(1.2)\left[(0+0) + 2(17.28 + 25.92 + 25.92 + 17.28)\right]$$
$$= 104 \text{ unit}^2$$

x	2	3.2	4.4	5.6	6.8	8
y=f(x)	0	17.28	25.92	25.92	17.28	0

Application and Interpretation (HL only)

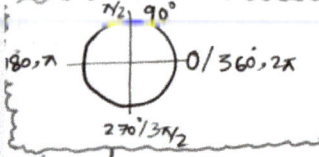

$2\sin x - \sqrt{3} = 0 \quad 0 \le x \le 2\pi$

$\sin x = \sqrt{3}/2$

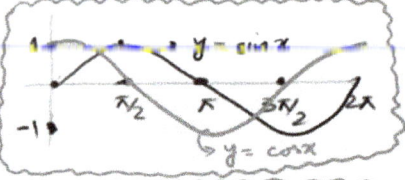

$x = \pi/6 , 5\pi/6$

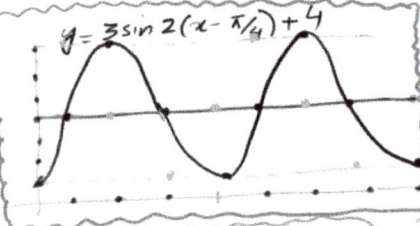

If $r = l$

1 radian

$l = r\theta$ $\quad A = \frac{1}{2} r^2 \theta$

$\nabla = \nabla - \nabla$

If $\sin\theta = 4/5$
and θ is obtuse
then $\cos\theta = ?$
$\tan\theta = ?$
then $\cos\theta = \frac{-3}{5}$
$\tan\theta = 5/-3$

$y = 3\sin 2(x - \pi/4) + 4$

$\sin 4\theta = 2\sin 2\theta \cos 2\theta$
$\sin 8\theta = 2\sin 4\theta \cos 4\theta$
$\cos 4\theta = \cos^2(2\theta) - \sin^2(2\theta)$

Degree $\to$ Radian
$70° \to 70 \times \frac{\pi}{180°}$

$m = \tan\theta = y = \tan\theta \cdot x + C$

TRIG - IDENTITIES

Radian $\to$ Degree
$3.4 \to 3.4 \times \frac{180}{\pi}$

$\tan\theta = \sin\theta/\cos\theta$

$\sin^2\theta + \cos^2\theta = 1$

$\sin 2\theta = 2\sin\theta\cos\theta$
$\cos 2\theta = \cos^2\theta - \sin^2\theta$
$\qquad = 2\cos^2\theta - 1$
$\qquad = 1 - 2\sin^2\theta$

Transition matrix

$T = \begin{bmatrix} 0.75 & 0.15 \\ 0.25 & 0.85 \end{bmatrix}$

Market shares (example from Revision Village)

Market share after 5 years

$T^5 S_0 = \begin{bmatrix} 0.75 & 0.15 \\ 0.25 & 0.85 \end{bmatrix}^5 \begin{bmatrix} 1 \\ 0 \end{bmatrix}$

$= \begin{bmatrix} 0.492 \\ 0.5764 \end{bmatrix}$ by GDC

Reflection in line
$y = (\tan a)x$

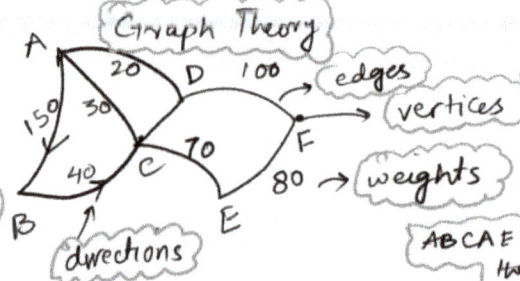

Graph Theory

$\to$ edges
$\to$ vertices
$\to$ weights

directions

Hamiltonian Path

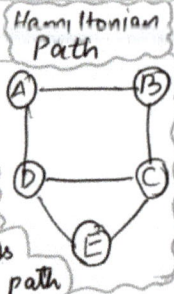

ABCAE is the path

Minimum Spanning tree

Prim's or Kruskals Algorithm

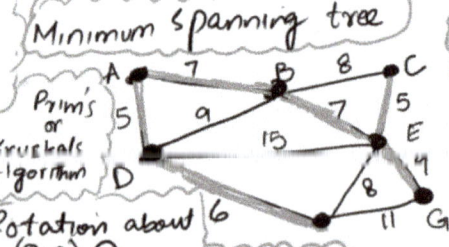

Rotation about (0,0)
$\begin{bmatrix} a & b \\ c & d \end{bmatrix} = \begin{bmatrix} \cos\theta & -\sin\theta \\ \sin\theta & \cos\theta \end{bmatrix}$

Geometric Transformation

$\begin{pmatrix} a & b \\ c & d \end{pmatrix}\begin{pmatrix} x \\ y \end{pmatrix} + \begin{pmatrix} h \\ k \end{pmatrix} = \begin{pmatrix} x' \\ y' \end{pmatrix}$

$\begin{bmatrix} a & b \\ c & d \end{bmatrix} = \begin{bmatrix} \cos 2\alpha & \sin 2\alpha \\ \sin 2\alpha & -\cos 2\alpha \end{bmatrix}$

Eulerian cycle

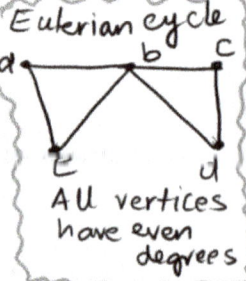

All vertices have even degrees

Translation By vector

$\begin{pmatrix} h \\ k \end{pmatrix}$

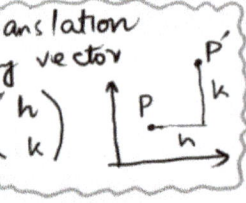

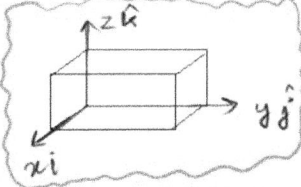

$$U = x\hat{i} + y\hat{j} + z\hat{k}$$

$$|U| = \sqrt{x^2 + y^2 + z^2}$$

$$\hat{U} = \frac{U}{|U|}$$

Vector Product

$$U = \begin{pmatrix} 4 \\ 5 \\ -4 \end{pmatrix} \qquad V = \begin{pmatrix} 3 \\ 2 \\ 1 \end{pmatrix}$$

$$U \times V = \begin{vmatrix} i & j & k \\ 4 & 5 & -4 \\ 3 & 2 & 1 \end{vmatrix}$$

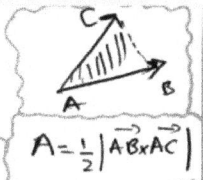

$$A = \frac{1}{2}|\vec{AB} \times \vec{AC}|$$

parallelogram law of addition of vectors

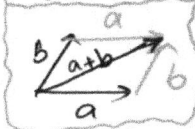

scalar product

$$U = \begin{pmatrix} U_1 \\ U_2 \\ U_3 \end{pmatrix} \qquad V = \begin{pmatrix} V_1 \\ V_2 \\ V_3 \end{pmatrix}$$

$$U \cdot V = \begin{pmatrix} U_1 V_1 \\ U_2 V_2 \\ U_3 V_3 \end{pmatrix}$$

$$U \cdot V = |U| \times |V| \times \cos\theta$$

$$\therefore \cos\theta = \frac{U \cdot V}{|U| \times |V|}$$

Vector equation of line

$$r = a + tb$$
Position vector, direction vector

$$U = 2\hat{i} + 4\hat{j} - 5k$$
$$V = 3i - 7\hat{j} - 8\hat{k}$$

$$U \cdot V = \begin{pmatrix} 2 \\ 4 \\ -5 \end{pmatrix} \cdot \begin{pmatrix} 3 \\ -7 \\ -8 \end{pmatrix} = 6 - 28 + 40 = 18$$

If $U \cdot V > 0$ acute angle
$U \cdot V = 0$ Right angle
$U \cdot V < 0$ Obtuse angle

$$\bar{x} - z\frac{\delta}{\sqrt{n}} < \mu < \bar{x} + z\frac{\delta}{\sqrt{n}}$$

confidence of Interval

90%
δ is known

$$\bar{x} - t\frac{S_{n-1}}{\sqrt{n}} < \mu < \bar{x} + t\frac{S_{n-1}}{\sqrt{n}}$$

$df = 12$

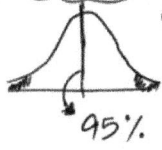

95%
δ is unknown

Quadratic Sine
Non-linear Regression by GDC
cubic Exponential logarithmic

$$X \sim N(\mu, \delta^2)$$

δ is unknown

$$t_{n-1} = \frac{\bar{x} - \mu}{\frac{S_{n-1}}{\sqrt{n}}}$$

δ is known

$$Z = \frac{\bar{X} - \mu}{\frac{\delta}{\sqrt{n}}}$$

Poisson Distribution

$$X \sim P_0(m)$$
$$P(X = x) = \frac{m^x e^{-m}}{x!}, \quad x = 0, 1, 2, 3 \cdots$$
$$E(X) = m$$
$$Var(X) = m$$
$$X \sim P_0(m_1) \qquad Y \sim P_0(m_2)$$
$$X + Y \sim P_0(m_1 + m_2)$$

Explains proportion of variability in y w.r.t x coefficient of Determination

$$R^2 = 1 - \frac{SS_{RES}}{SS_{TOT}}$$

$$R^2 = 1 \iff SS_{RES} = 0$$

$$\frac{dy}{dx} = \lim_{h \to 0} \frac{f(x+h) - f(x)}{h}$$

$$\frac{dy}{dx} = y' = f'(x)$$

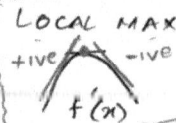

LOCAL MAX
+ive, −ive
$f'(x)$

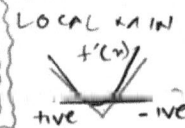

LOCAL MIN
$f'(x)$
+ive − ive

Graphical Interpretation of $y \to y' \to y''$

$$\frac{d}{dx}(C) = 0$$

$$\frac{d}{dx}(Cx) = C$$

$$\frac{d}{dx} x^n = n x^{n-1}$$

$$\frac{d}{dx}(UV) = UV' + VU'$$

$$\frac{d}{dV}\left(\frac{U}{V}\right) = \frac{VU' - UV'}{V^2}$$

$$\frac{d}{dx} \sin x = \cos x$$

$$\frac{d}{dx} \cos x = -\sin x$$

$$\frac{d}{dx} \ln|x| = \frac{1}{x}$$

$$\frac{d}{dx} e^x = e^x$$

$f'(x) = 0$

$f''(x) = \dfrac{d^2 y}{dx^2}$

point of inflexion

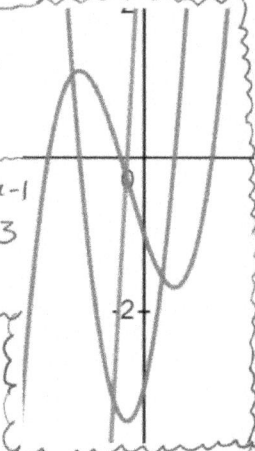

$y = 3x^3 + 2x^2 - 3x - 1$

$y' = 9x^2 + 4x - 3$

$y'' = 18x + 4$

$-2-$

CHAIN RULE

$$\frac{dy}{dx} = \frac{dy}{du} \times \frac{du}{dx}$$

Example $y = \ln \ln|x|$

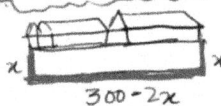
OPTIMIZATION

300 m rope
Area to max?
$A = x(300 - 2x)$
$A = 300x - 2x^2$
$A'(x) = 300 - 4x$
$A'(x) = 0 \Rightarrow x = 75$
$A(75) = 11250 \text{ m}^2$

x ⬚ x
$300 - 2x$

$$y = \ln(\ln|x|)$$
let $u = \ln x$ then $y = \ln U$
$du/dx = \frac{1}{x}$ $dy/du = \frac{1}{U}$

$$\frac{dy}{dx} = \frac{dy}{du} \times \frac{du}{dx}$$
$$= \frac{1}{U} \times \frac{1}{x}$$
$$= \frac{1}{\ln x} \times \frac{1}{x}$$

$$\int 1 \, dx = x + c$$

$$\int x^n dx = \frac{x^{n+1}}{n+1} + C$$

$$\int \cos x \, dx = \sin x + c$$

$$\int \sin x \, dx = -\cos x + c$$

$$\int e^x = e^x + C$$

$$\int \frac{1}{x} dx = \ln|x| + c$$

Area Under Curve

$y = \frac{1}{2}x^2 + 2$

$$\int_1^3 \left(\frac{1}{2}x^2 + 2\right) dx$$

$$\left[\frac{1}{2}\frac{x^3}{3} + 2x\right]_1^3 = 8.33 \text{ unit}^2$$

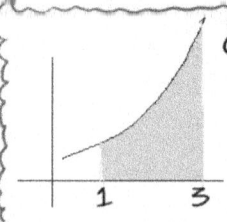

1   3

FURTHER INTEGRATION

$$\int \frac{f'(x)}{f(x)} dx = \ln|f(x)| + C \Rightarrow$$

$$\int \frac{1}{x \ln x} dx \Rightarrow \int \frac{\frac{1}{x}}{\ln x} dx$$
$$= \ln \ln|x| + c$$

KINEMATICS

S = displacement $\int s \, dt$
$\frac{ds}{dt}$
V = velocity $\int dt$
$\frac{dv}{dt}$
a = acceleration

BY SUBSTITUTION

$$\int x(2x^2 + 8)^8 dx \Rightarrow$$

$U = 2x^2 + 8$
$du = 4x \, dx$
$\frac{du}{4} = x \, dx$

$$\int U^8 \cdot \frac{du}{4} = \frac{1}{4}\int U^8 du$$
$$= \frac{1}{4} \frac{U^{8+1}}{(8+1)} = \frac{1}{36}(2x^2 + 8)^9 + C$$

$$\int x^2 e^{4x^3 + 3} dx = \frac{1}{12}\int e^U du$$

$U = 4x^3 + 3$
$du = 12x^2 dx$
$\frac{du}{12} = x^2 dx$

$$= \frac{1}{12} e^U$$
$$= \frac{1}{12} e^{4x^3 + 3} + C$$

Calculus

Volume of Revolution

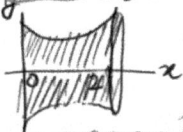

$360°$ or 2π about x-axis

$$V = \int_a^b \pi \left[f(x)\right]^2 dx$$

$f(x) = 3x^3 + 2x^2 - 3x + 4$

$$\text{Volume} = \pi \int_0^2 (3x^3 + 2x^2 - 3x + 4)^2 dx$$

$$= 947 \text{ unit}^3$$

$360°$ or 2π about x-axis

$$V = \int_a^b \pi \, x^2 \, dy$$

$$\frac{dy}{dx} = f(x,y) \quad \text{First Order}$$

Second order
↓
Euler's Method | **Differential Equation**

$$y_{n+1} = y_n + h \times f(x_n, y_n);$$
$$x_{n+1} = x_n + h \ (\text{step length})$$

For coupled Equations
$$x_{n+1} = x_n + h \times f(x_n, y_n, t_n)$$

$$t_{n+1} = t_n + h$$
$$y_{n+1} = y_n + f_2(x_n, y_n, t_n)$$

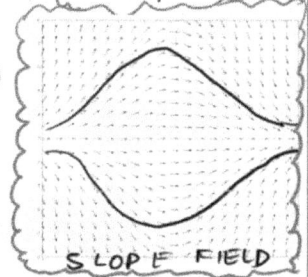

SLOPE FIELD

Analysis and Approaches (SL)

Number and Algebra

$$S_\infty = 80 + 40 + 20 + 10 + 5 + \cdots = \frac{80}{1 - 0.5} = 160$$
$$u_1 = 80, \quad r = 0.5$$

INFINITE GEOMETRIC SEQUENCE
when $0 < |r| < 1$
$$S_\infty = u_1 + u_1 r + u_1 r^2 + u_1 r^3 + \cdots = \frac{u_1}{1 - r}$$

PASCAL'S Δ

```
      1
    1   1
   1  2  1
  1  3  3  1
 1  4  6  4  1
1  5  10 10 5  1
      5C₃
```

$$^nC_r = \frac{n!}{r!(n-r)!}$$
where $n! = n \times (n-1) \times (n-2) \times \cdots 3 \cdot 2 \cdot 1$

BINOMIAL THEOREM

FINDING n^{th} Term
$$\binom{n}{r} a^{n-r} b^r$$

$$(a+b)^n = \,^nC_0 a^n b^0 + \,^nC_1 a^{n-1} b^1 + \,^nC_2 a^{n-2} b^2 \cdots \,^nC_n a^0 b^n$$

LAWS OF LOGARITHM
$$\log_a xy = \log_a x + \log_a y$$
$$\log_a \left(\frac{x}{y}\right) = \log_a x - \log_a y$$
$$\log_a x^n = n \log_a x$$
$$7^{\log_7 100} = 100$$

LAWS OF EXPONENTS / RATIONAL EXPONENTS
$$a^m \cdot a^n = a^{m+n}$$
$$\frac{a^m}{a^n} = a^{m-n}$$
$$(a^m)^n = a^{mn}$$

$$a^{1/m} = \sqrt[m]{a}$$
$$a^{m/n} = \sqrt[n]{a^m} \text{ or}$$
$$a^{n/m} = \sqrt[m]{a^n}$$
$$a^{-m} = \frac{1}{a^m}$$

CHANGE OF BASE RULE

$$\log_{10} 100 = \frac{\log_2 100}{\log_2 10} = \frac{\log_5 100}{\log_5 10}$$

$$\log_2(32) = \log_2 2^5 = 5\log_2 2 = 5$$
$$\log_2(14) = \log_2 7 + \log_2 2$$
$$\log_2\left(\frac{12}{5}\right) = \log_2 12 - \log_2 5$$

Functions

Functions | input → process → output
x^2 $f(x) = y$

LINEAR FUNCTION

$ax + by + c = 0$
$y = mx + c$
$y - y_1 = m(x - x_1)$

$f(x) = ax^2 + bx + c$ (standard form)

$a > 0$ $\smile$
$a < 0$ $\frown$

QUADRATICS

$f(x) = a(x-h)^2 + k$
VERTEX FORM (h, k)

y-intercept $x = 0$

x-intercepts $y = 0$

$f(x) = a(x-p)(x-q)$
FACTOR FORM

$(p, 0)$ $(0, c)$ $(q, 0)$

DISCRIMINANT
$\Delta = b^2 - 4ac$

$\Delta < 0$ no real root

$\Delta = 0$ Repeated root

$\Delta > 0$ Two distinct Real roots

axis of symmetry
$x = -b/2a$

$y = 2$
$(0, 0.67)$
$x = 3/2$
$(0.5, 0)$

RATIONAL FUNCTIONS
$f(x) = \dfrac{4x-2}{2x-3}$

Horizontal asymptote
as $x \to 0$ then $y = 2$

vertical asymptote
$2x - 3 = 0$
$2x = 3$
$x = 3/2$

EXPONENTIAL AND LOGARITHMIC FUNCTIONS

$y = a^x$ $y = x$

$y = a^x$
Inverses
$y = \log_a x$

COMPOSITE FUNCTION

$x \to g(x) \to f(g(x))$
   g f

$f \circ g \circ x = f(g(x))$

$\begin{pmatrix} 2 \\ 3 \end{pmatrix} \to \begin{pmatrix} 4 \\ 9 \end{pmatrix}$ one to one

$\begin{pmatrix} 2 \\ -2 \end{pmatrix} \to \begin{pmatrix} 4 \end{pmatrix}$ Many to one

$f \circ f^{-1}(x) = f^{-1} \circ f(x) = x$

$a f(x)$

TRANSFORMATION
$-f(x)$

$f(x-h) + k$

Geometry and Trigonometry

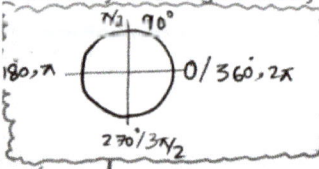

$\pi/2$, $90°$
180, π $0 / 360$, 2π
$270° / 3\pi/2$

if $r = l$
r r → 1 radian

$l = r\theta$ $A = \frac{1}{2} r^2 \theta$

$\bigcirc = \bigvee - \bigvee$

Degree → Radian
$70° \to 70 \times \dfrac{\pi}{180°}$

Radian → Degree
$3.4 \to 3.4 \times \dfrac{180}{\pi}$

$2 \sin x - \sqrt{3} = 0$ $0 \leq x \leq 2\pi$
$\sin x = \sqrt{3}/2$

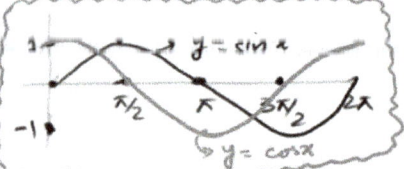

$5\pi/6$
S | A
T | C

$x = \pi/6 , 5\pi/6$

If $\sin\theta = 4/5$
and θ is obtuse
then $\cos\theta = ?$
   $\tan\theta = ?$
then $\cos\theta = \dfrac{-3}{5}$
   $\tan\theta = 5/-3$

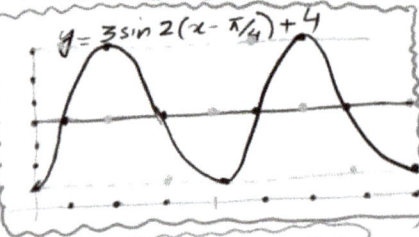

4 | 5
-3

$m = \tan\theta = y = \tan\theta \cdot x + c$

TRIG - IDENTITIES

$\tan\theta = \sin\theta / \cos\theta$

$\sin^2\theta + \cos^2\theta = 1$

$y = \sin x$
$\pi/2$ π $3\pi/2$ 2π
-1
$y = \cos x$

$y = 3\sin 2(x - \pi/4) + 4$

$\sin 4\theta = 2 \sin 2\theta \cos 2\theta$
$\sin 8\theta = 2 \sin 4\theta \cos 4\theta$
$\cos 4\theta = \cos^2(2\theta) - \sin^2(2\theta)$

$\sin 2\theta = 2\sin\theta\cos\theta$
$\cos 2\theta = \cos^2\theta - \sin^2\theta$
$= 2\cos^2\theta - 1$
$= 1 - 2\sin^2\theta$

Mutually Exclusive Event

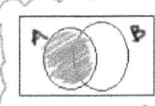

$$P(A \cup B) = P(A) + P(B)$$

Combined Events

$A \cap B$

$$P(A \cup B) = P(A) + P(B) - P(A \cap B)$$

→ $A \cap B'$

→ $A' \cap B$

32 students
19 takes Bio
23 Chem
3 Neither

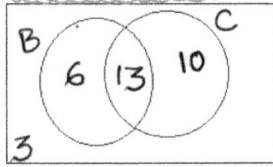

B ⬡ C

6 | 13 | 10

3

$$32 - 3 = 29 \neq 23 + 19$$

Linear Regression

Can be used to predict y for given x, cannot be used x for given y

$$y = mx + c$$

x

$$E(X) = np \quad V(X) = np(1-p)$$

CONDITIONAL PROBABILITY

$$P(A/B) = \frac{P(A \cap B)}{P(B)}$$

$$P(C/B') = \frac{10}{13}$$

If C and B are independent?

$$P(C) \cdot P(B) = P(C \cap B)$$

$$\frac{23}{32} \times \frac{19}{32} \neq \frac{13}{32}$$

So they are independent.

μ 75%

STANDARDISED NORMAL VARIABLE

z repesents how many standard deviation X is away from its mean

80 100

$\delta = ?$

$Z \sim N(0,1)$

$$Z = \frac{X - \mu}{\sigma}$$

$$-0.674 = \frac{80 - 100}{\sigma}$$

$$\sigma = 29.7$$

Calculus

$$\frac{dy}{dx} = \lim_{h \to 0} \frac{f(x+h) - f(x)}{h}$$

$$\frac{dy}{dx} = y' = f'(x)$$

$$\frac{d}{dx}(C) = 0$$

$$\frac{d}{dx}(Cx) = C$$

$$\frac{d}{dx} x^n = n x^{n-1}$$

$$\frac{d}{dx}(UV) = UV' + VU'$$

$$\frac{d}{dv}\left(\frac{U}{V}\right) = \frac{VU' - UV'}{V^2}$$

$$\frac{d}{dx} \sin x = \cos x$$

$$\frac{d}{dx} \cos x = -\sin x$$

$$\frac{d}{dx} \ln|x| = \frac{1}{x}$$

$$\frac{d}{dx} e^x = e^x$$

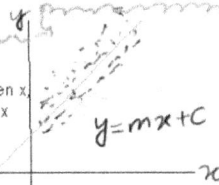

LOCAL MAX
+ive -ive
$f'(x)$

LOCAL MIN
$f'(x)$
+ive -ive

$f''(x) = 0$

point of inflexion

$$f''(x) = \frac{d^2y}{dx^2}$$

Graphical Interpretation of $y \to y' \to y''$

$$y = 3x^3 + 2x^2 - 3x - 1$$
$$y' = 9x^2 + 4x - 3$$
$$y'' = 18x + 4$$

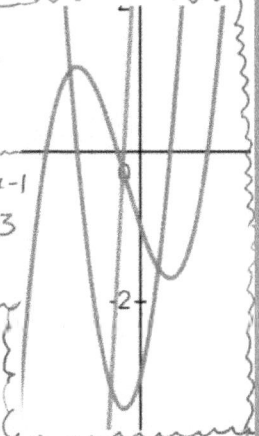

-2

CHAIN RULE

$$\frac{dy}{dx} = \frac{dy}{du} \times \frac{du}{dx}$$

Example $y = \ln \ln|x|$ →

OPTIMIZATION

300 m rope
Area to max?
$$A = x(300 - 2x)$$
$$A = 300x - 2x^2$$
$$A'(x) = 300 - 4x$$
$$A'(x) = 0 \implies x = 75$$
$$A(75) = 11250 \text{ m}^2$$

$$y = \ln(\ln|x|)$$

let $U = \ln x$ then $y = \ln U$
$$\frac{du}{dx} = \frac{1}{x} \quad \frac{dy}{du} = \frac{1}{U}$$

$$\therefore \frac{dy}{dx} = \frac{dy}{du} \times \frac{du}{dx}$$
$$= \frac{1}{U} \times \frac{1}{x}$$
$$= \frac{1}{\ln x} \times \frac{1}{x}$$

Calculus

$$\int 1\,dx = x + c \qquad\qquad \int \sin x\,dx = -\cos x + c$$

$$\int x^n\,dx = \frac{x^{n+1}}{n+1} + c \qquad \int e^x = e^x + c$$

$$\int \cos x\,dx = \sin x + c \qquad \int \frac{1}{x}\,dx = \ln|x| + c$$

Area Under Curve

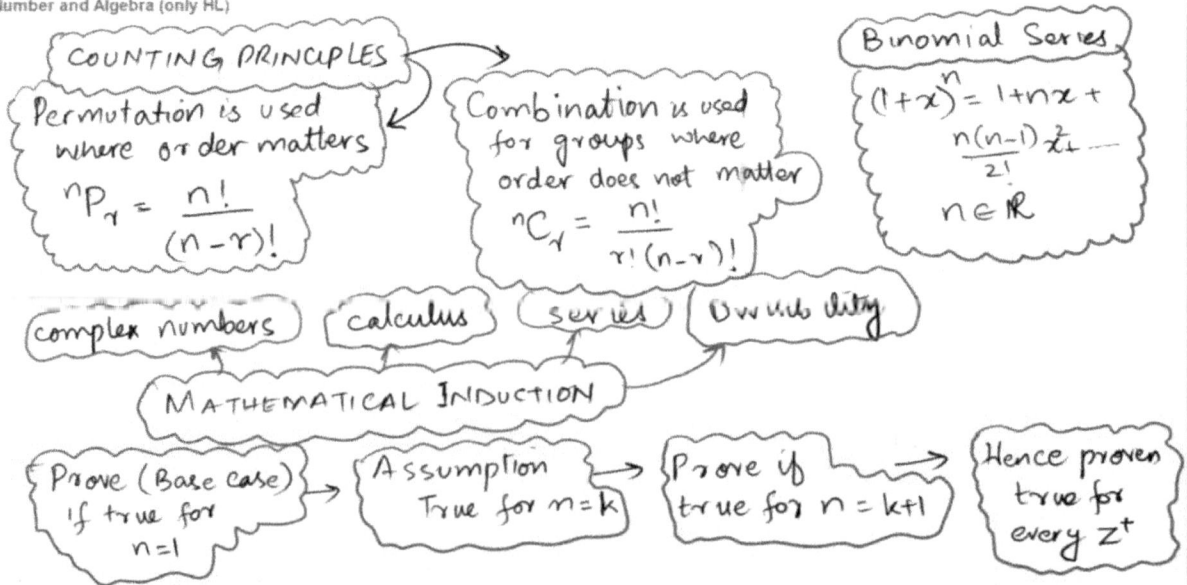

$$y = \frac{1}{2}x^2 + 2$$

$$\int_1^3 \left(\frac{1}{2}x^2 + 2\right)dx$$

$$\left[\frac{1}{2}\frac{x^3}{3} + 2x\right]_1^3 = 8.33 \; \text{unit}^2$$

FURTHER INTEGRATION

$$\int \frac{f'(x)}{f(x)}\,dx = \ln|f(x)| + c \;\Rightarrow$$

$$\int \frac{1}{x\ln x}\,dx \Rightarrow \int \frac{\frac{1}{x}}{\ln x}\,dx$$

$$= \ln \ln|x| + c$$

KINEMATICS

$$\frac{ds}{dt} \quad S = \text{displacement} \quad \int s\,dt$$

$$\quad V = \text{velocity}$$

$$\frac{dv}{dt} \quad a = \text{acceleration} \quad \int dt$$

BY SUBSTITUTION

$$\int x(2x^2 + 8)^8\,dx \;\Rightarrow \qquad U = 2x^2 + 8$$
$$du = 4x\,dx$$
$$\int u^8 \cdot \frac{du}{4} = \frac{1}{4}\int u^8\,du \qquad \frac{du}{4} = x\,dx$$
$$= \frac{1}{4}\frac{U^{8+1}}{(8+1)} = \frac{1}{36}(2x^2+8)^9 + c$$

$$\int x^2 e^{4x^3+3}\,dx = \frac{1}{12}\int e^u\,du$$
$$U = 4x^3 + 3$$
$$du = 12x^2\,dx = \frac{1}{12}e^u$$
$$\frac{du}{12} = x^2\,dx = \frac{1}{12}e^{4x^3+3} + c$$

Number and Algebra (only HL)

COUNTING PRINCIPLES

Permutation is used where order matters

$$^nP_r = \frac{n!}{(n-r)!}$$

Combination is used for groups where order does not matter

$$^nC_r = \frac{n!}{r!(n-r)!}$$

Binomial Series

$$(1+x)^n = 1 + nx + \frac{n(n-1)}{2!}x^2 \ldots$$

$$n \in \mathbb{R}$$

complex numbers   calculus   series   Divisibility

MATHEMATICAL INDUCTION

Prove (Base case) if true for n=1 → Assumption True for n=k → Prove if true for n=k+1 → Hence proven true for every Z^+

Analysis and Approaches (HL)

real, imaginary

$$z = a + ib$$

CARTESIAN FORM

$$r^2 = a^2 + b^2$$

$$\theta = \tan^{-1}\left(\frac{b}{a}\right)$$

$$z = r e^{i\theta}$$
Euler's form

De Moivre's Theorem
$$(r\,cis\,\theta)^n = r^n cis\,n\theta$$

SYSTEM OF LINEAR EQUATIONS

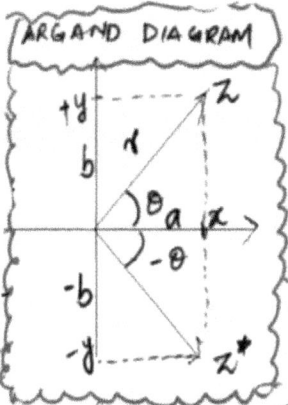

ARGAND DIAGRAM

Mod $z = |z| = r$
Arg $z = \theta$

COMPLEX NUMBERS

$$z = r\,cis\,\theta$$
$$z = r(\cos\theta + i\sin\theta)$$
Mod- Argument form
polar form

Product of $z_1 z_2$
$z_1 = 5\,cis\,\frac{\pi}{2}$ $z_2 = 3\,cis\,\frac{\pi}{3}$
$$z_1 z_2 = 15\,cis\left(\frac{\pi}{2} + \frac{\pi}{3}\right)$$

Quotient of $\frac{z_1}{z_2}$
$$= \frac{5\,cis\,\frac{\pi}{2}}{3\,cis\,\frac{\pi}{3}}$$
$$= \frac{5}{3}\,cis\left(\frac{\pi}{2} - \frac{\pi}{3}\right)$$

n^{th} Root of complex NO
$$z = 64^{1/3}\left(\cos\pi + i\sin\pi\right)^{1/3}$$
$$= 4\left[\cos\left(\frac{\pi + 2k\pi}{3}\right) + i\sin\left(\frac{\pi + 2k\pi}{3}\right)\right]$$

$$3x + 4y - 9z = 15$$
$$4x - 7y + 10z = 7$$
$$13x + 8y \cdots$$

General solution
No Solution
Infinite sol
Unique Sol

isistant

$$P(x) = x^2 - Sum(x) + Product$$

POLYNOMIALS

If $f(a) = 0$ then
$(x-a)$ is a factor

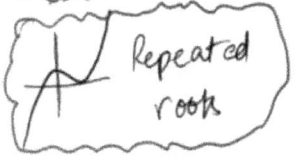

2R, 2C,

Repeated roots

Even function

$$ax^2 + bx + c = 0$$
if α, β are two roots
$$\alpha + \beta = -b/a$$
$$\alpha\beta = \frac{c}{a}$$

MODULUS
$$|x| = x \text{ if } x \geq 0$$
$$|x| = -x \text{ if } x < 0$$

$$f(-x) = f(x)$$
eg $f(x) = x^2$
$$f(-x) = x^2$$

odd Function
$$f(-x) = -f(x)$$
$$f(x) = x^3$$
$$f(-x) = -x^3$$

$$ax^3 + bx^2 + cx + d = 0$$
$$\alpha + \beta + \gamma = \frac{-b}{a}$$
$$\alpha\beta\gamma = \frac{-d}{a}$$

Vieta's Rule
$$Sum = \frac{-a_{n-1}}{a_n}$$
$$Product = \frac{(-1)^n a_0}{a_1}$$

$sec x = \dfrac{1}{\cos x}$

$cosec x = \dfrac{1}{\sin x}$ Reciprocal Trigonometric Ratios

$cot x = \dfrac{1}{\tan x}$

$1 + \tan^2\theta = \sec^2\theta$

$1 + \cot^2\theta = \csc^2\theta$

COMPOUND ANGLE IDENTITY

$\sin(\alpha \pm \beta) = \sin\alpha\cos\beta \pm \cos\alpha\sin\beta$

$\cos(\alpha \pm \beta) = \cos\alpha\cos\beta \mp \sin\alpha\sin\beta$

$\tan(\alpha \pm \beta) = \dfrac{\tan\alpha \pm \tan\beta}{1 \mp \tan\alpha\tan\beta}$

$f(x)$ is a continuous at $x = c$ if

$\lim\limits_{x \to c} f(x) = f(c)$ within domain

LIMITS AND CONTINUITY

VERTICAL ASYMPTOTE

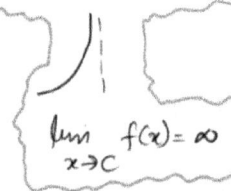

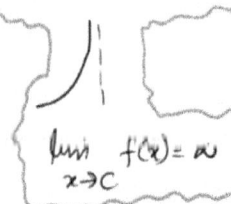

$\lim\limits_{x \to c} f(x) = \infty$

HORIZONTAL ASYMPTOTE

$\lim\limits_{x \to \infty} f(x) = k$

SANDWITCH THEOREM

$\lim\limits_{\theta \to 0} \dfrac{\sin\theta}{\theta} = 1$

Geometry and Trigonomtery (only HL)

$sec x = \dfrac{1}{\cos x}$

$cosec x = \dfrac{1}{\sin x}$ Reciprocal Trigonometric Ratios

$cot x = \dfrac{1}{\tan x}$

$1 + \tan^2\theta = \sec^2\theta$

$1 + \cot^2\theta = \csc^2\theta$

COMPOUND ANGLE IDENTITY

$\sin(\alpha \pm \beta) = \sin\alpha\cos\beta \pm \cos\alpha\sin\beta$

$\cos(\alpha \pm \beta) = \cos\alpha\cos\beta \mp \sin\alpha\sin\beta$

$\tan(\alpha \pm \beta) = \dfrac{\tan\alpha \pm \tan\beta}{1 \mp \tan\alpha\tan\beta}$

$f(x)$ is a continuous at $x = c$ if

$\lim\limits_{x \to c} f(x) = f(c)$ within domain

LIMITS AND CONTINUITY

VERTICAL ASYMPTOTE

$\lim\limits_{x \to c} f(x) = \infty$

HORIZONTAL ASYMPTOTE

$\lim\limits_{x \to \infty} f(x) = k$

SANDWITCH THEOREM

$\lim\limits_{\theta \to 0} \dfrac{\sin\theta}{\theta} = 1$

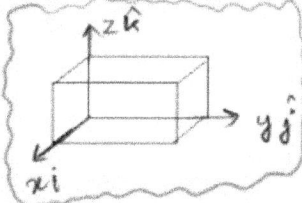

$$U = x\hat{i} + y\hat{j} + z\hat{k}$$

$$|U| = \sqrt{x^2 + y^2 + z^2}$$

$$\hat{U} = \frac{U}{|U|}$$

Vector Product

$$U = \begin{pmatrix} 4 \\ 5 \\ -4 \end{pmatrix} \quad V = \begin{pmatrix} 3 \\ 2 \\ 1 \end{pmatrix}$$

$$U \times V = \begin{vmatrix} i & j & k \\ 4 & 5 & -4 \\ 3 & 2 & 1 \end{vmatrix}$$

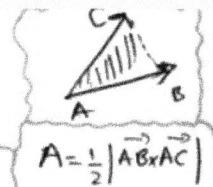

$$A = \frac{1}{2} |\overrightarrow{AB} \times \overrightarrow{AC}|$$

parallelogram law of addition of vectors

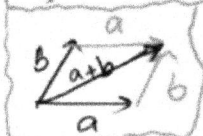

Scalar product

$$U = \begin{pmatrix} U_1 \\ U_2 \\ U_3 \end{pmatrix} \quad V = \begin{pmatrix} V_1 \\ V_2 \\ V_3 \end{pmatrix}$$

$$U \cdot V = \begin{pmatrix} U_1 V_1 \\ U_2 V_2 \\ U_3 V_3 \end{pmatrix}$$

$$U \cdot V = |U| \times |V| \times \cos\theta$$

$$\therefore \cos\theta = \frac{U \cdot V}{|U| \times |V|}$$

Vector equation of line

$$r = a + tb$$

Position vector / direction vector

$$U = 2\hat{i} + 4\hat{j} - 5k$$

$$V = 3i - 7\hat{j} - 8\hat{k}$$

$$U \cdot V = \begin{bmatrix} 2 \\ 4 \\ -5 \end{bmatrix} \cdot \begin{bmatrix} 3 \\ -7 \\ -8 \end{bmatrix} = 6 - 28 + 40 = 18$$

If $U \cdot V > 0$ acute angle

$U \cdot V = 0$ Right angle

$U \cdot V < 0$ Obtuse angle

Equation of Plane using Normal Vector

$$r \cdot n = a \cdot n$$

Cartesian Equation of a plane

$$ax + by + cz = d$$

$$P(B_i/A) = \frac{P(B_i) \cdot P(A/B_i)}{P(B_1)P(A/B_1) + P(B_2)P(A/B_2) + P(B_3)P(A/B_3)}$$

THREE EVENTS

BAY'S THEOREM

TWO EVENTS

$$P(B/A) = \frac{P(B) \cdot P(A/B)}{P(B)P(A/B) + P(B')P(A/B')}$$

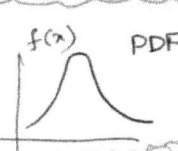

$f(x)$ PDF

$$\int_{-\infty}^{\infty} f(x)dx = 1$$

$$\mu = E(X) = \int_{-\infty}^{\infty} x\, f(x)dx$$

PROBABILITY DISTRIBUTION FUNCTION

MODE

Max of $f(x)$ is mode

Median $= m$

$$\int_{-\infty}^{m} f(x)dx = \frac{1}{2}$$

$$Var(X) = E(x^2) - \{E(X)\}^2$$
$$= \int_{-\infty}^{\infty} x^2 f(x)dx - \mu^2$$

SOME MORE DERIVATIVES

$\frac{d}{dx} \tan x = \sec^2 x$

$\frac{d}{dx} (\sec x) = \sec x \tan x$

$\frac{d}{dx} (\csc x) = -\csc x \cot x$

$\frac{d}{dx} (\cot x) = -\csc^2 x$

$\frac{d}{dx} (a^x) = a^x \cdot \ln a$

$\frac{d}{dx} \log_a x = \frac{1}{x \cdot \ln a}$

$\frac{d}{dx} \sin^{-1} x = \frac{1}{\sqrt{1-x^2}}$

$\frac{d}{dx} \cos^{-1}(x) = -\frac{1}{\sqrt{1-x^2}}$

$\frac{d}{dx} \tan^{-1} x = \frac{1}{1+x^2}$

SOME MORE INTEGRALS

$\int a^x \, dx = \frac{1}{\ln a} \cdot a^x + C$

$\int \frac{1}{a^2+x^2} \, dx = \frac{1}{a} \tan^{-1}\left(\frac{x}{a}\right) + C$

$\int \frac{1}{\sqrt{a^2-x^2}} \, dx = \sin^{-1}\left(\frac{x}{a}\right) + C$

Integration by Parts

$\int u \frac{dv}{dx} \, dx = UV - \int v \frac{du}{dx} \cdot dx = \int u \, dv = uv - \int v \, du$

$I = \int \sin x \cdot e^x \, dx$ (Repeated Integrals)

$= \sin x \cdot e^x - \int \cos x \cdot e^x \, dx$

$= \sin x \cdot e^x - \left[\cos x \cdot e^x - \int -\sin x \, e^x \, dx\right]$

$I = \sin x \cdot e^x - e^x \cos x - I \Rightarrow I = \frac{e^x}{2}(\sin x - \cos x)$

eg $\int \ln x \cdot 1$

$= \ln x \cdot x - \int \frac{1}{x} \cdot x \, dx$

$= x \cdot \ln x - x + C$

Differential Equation

EULER'S METHOD $\Rightarrow y_{n+1} = y_n + h f(x_n, y_n)$ where $x_{n+1} = x_n + h$

Variable Seperable

$\frac{dy}{dx} = 2x$

$\int dy = 2\int x \, dx$

$y = 2\frac{x^2}{2} + C$

$y = x^2 + C$

Maclaurin Series

$f(x) = f(0) + x f'(0) + \frac{x^2}{2!} f''(0)$

$+ \cdots$

$e^x = 1 + x + \frac{x^2}{2!} + \cdots$

Homogeneous form

$\frac{dy}{dx} = f(y/x)$

Sub $y = vx \Rightarrow \frac{dy}{dx} = \frac{x \, dv}{dx} + v$

$\frac{dy}{dx} = \frac{x+2y}{x} = 1 + 2\left(\frac{y}{x}\right)$

$\therefore x\frac{dv}{dx} + v = 1 + 2\left(\frac{vx}{x}\right)$

$x\frac{dy}{dx} = 1 + V$

$\int \frac{1}{1+v} \, dv = \int \frac{1}{x} \, dx$

$\ln|1+v| = \ln|x| + C$

$y = Ax^2 - x$

Integrating factor

$\frac{dy}{dx} + P(x)y = Q(x)$

Multiply

$I(x) = e^{\int P(x) \, dx}$

$\frac{dy}{dx} + 2y = e^x$

$e^{2x}\frac{dy}{dx} + 2e^{2x}y = e^{3x}$

$\int \frac{dy}{dx} y \cdot e^{2x} = \int e^{3x}$

$y e^{2x} = \frac{1}{3} e^{3x} + C$

$y = \frac{e^x}{3} + \frac{C}{e^{2x}}$

Understanding the IA Criterions and thoroughly understanding their ingredients.

Here following points should be kept scoring optimal in each criterion. The examples for each criterion have been mentioned for each criterion.

Presentation criterion A:

0 The exploration does not reach the standard described by the descriptors below.

1 The exploration has some coherence or some organization.

2 The exploration has some coherence and shows some organization.

3 The exploration is coherence and well organized.

 4 The exploration is coherent, well organized, and concise.

Presentation is one of the most essential components of the IA as it overall develops an impression on the examiner. To score optimal in this criterion following things should be kept in mind as this what an examiner mainly looks for.

- ✓ Front page should have the title and number of pages. Here are the few examples.

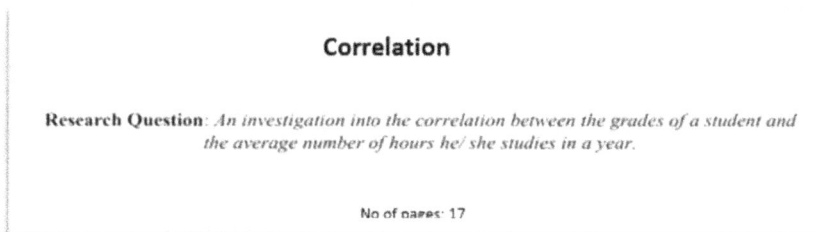

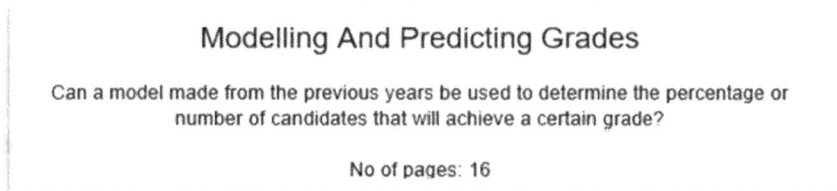

Correlation

Research Question: *An investigation into the correlation between the grades of a student and the average number of hours he/she studies in a year.*

No of pages: 17

Modelling And Predicting Grades

Can a model made from the previous years be used to determine the percentage or number of candidates that will achieve a certain grade?

No of pages: 16

- ✓ 12-20 pages with double line spacing
- ✓ Page numbering
- ✓ Introduction
- ✓ Specific Aim: In general, presentation means the overall organization and coherence of the IA. If your classmate can understand and follow your IA the way you understand it, it means, your IA is well presented. Being an examiner, we look for three things in the introduction, i.e., aim, rationale and plan. Given below are the two examples for the clear aim and an unclear aim in the IA.

✓ Focused Rationale

✓ Transparent Plan. There should be a clear approach and plan given in the introduction as mentioned below

Plan and Approach

To answer this question only secondary data could be used which was provided in the IB statistical bulletins that are present for anyone to view.

My first instinct was to take the data from the last 5 years and take the averages of the data after which I would plot it on a cumulative frequency curve but there were a few problems with that. Firstly, the May and November examinations had slightly different pass rates and mean grades which made it so that an average of both the sessions wouldn't be able to capture the most accurate picture of the last five years.

The second problem I found was that I needed to choose whether to plot the kids who didn't receive the diploma or the ones that did because in the statistical bulletins the percentages of each are different. Since all my seniors obtained the Diploma, I decided that the latter was a better option.

Listening to their grades the expected cumulative frequency curve would follow a logarithmic curve of $y = alnx + b$ with a vertical stretch factor of a and a y-intercept of b.

To test my hypothesis, I decided to take a larger data set and then use that to see if it would follow the proposed path as more people are near the middle leading to a steeper gradient at the start which slowly reduces to become an asymptote at 100% in an ideal case.

- ✓ The concepts should be elaborated clearly (avoid leaving the reader in the grey area)
- ✓ The transition between the topics and distinct concept should be linked (Fluidity of the document should be clear). Before starting a new paragraph, make your reader prepared what is going to come in the next paragraph and how it is related to the previous one by providing a rationale behind it.
- ✓ Avoid adding long table (Prefer appendices)
- ✓ Avoid repetition: This happens mostly in the IA and students do not take this into account and lose score in lacking conciseness of the IA.

Conclusion

It can be deduced that the data does follow a normal distribution however the scores of the students that have obtained the Diploma lie well in between three standard deviations of the mean and so the third empirical rule doesn't hold a lot of value in this case. Any how it came as a surprise that the point distribution does follow somewhat of a normal distribution and thus using the normal CDF function I will try and answer my question of: was the grade distribution for the May 2020 exams like previous years? And how closely does it resemble it.

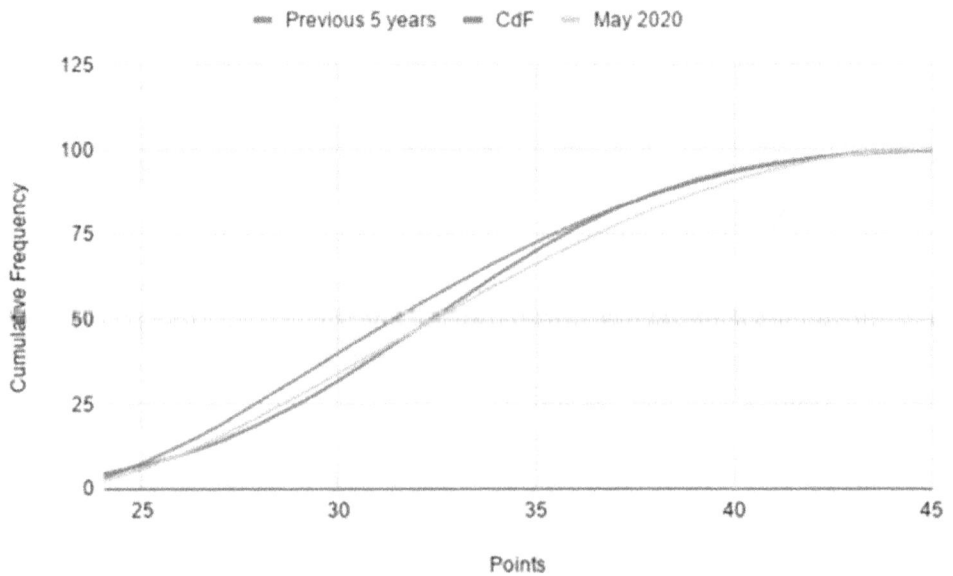

Fig. 10 CDF compared to the May 2020 and Previous 5 May examinations

This is a very interesting site to see. The CDF line follows the curve of the May 2020 examinations at the beginning and then merges with the line of the previous 5 May examinations. After seeing this I am confident that the grade distributions follow a normal distribution to a large extent and no specific points were awarded out of proportion to the May 2020 batch. Using excel I found the coefficient of determination r^2 of the Cdf with the previous 5 years to be 0.993465089 while for the May 2020 examinations it turned out to be 0.99616521. Although a minute difference it is safe to say that the CDF model is very precise at predicting the number of candidates receiving a particular grade.

Mathematical Communication B:

0 The exploration does not reach the standard described by the descriptors below.

1 The exploration contains some relevant mathematical communication which is partially appropriate.

2 The exploration contains some relevant mathematical communication.

3 The mathematical communication is relevant, appropriate, and mostly consistent.

4 The mathematical communication is relevant, appropriate, and consistent throughout.

Mathematical communication is one of the most technical aspects of the IA as the examiner takes it quite seriously. By following the checklist below once can ensure maximum marks in this criterion.

- ✓ Define all key terms and variables
- ✓ Use real arithmetic sign (avoid generic signs such as * for multiplication or / for divisor)
- ✓ Write equations through an equation editor (Don't use alphabets such as use $2x + y$ instead of 2x+y)

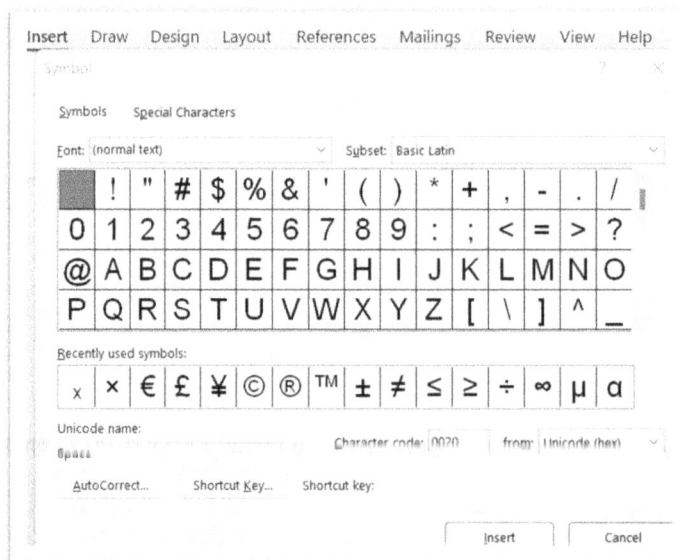

- ✓ Center align the equations
- ✓ Along with equations the chart, table, graphs etc. should also be aligned in center
- ✓ Axes should be clearly labelled along with clear title of each chart

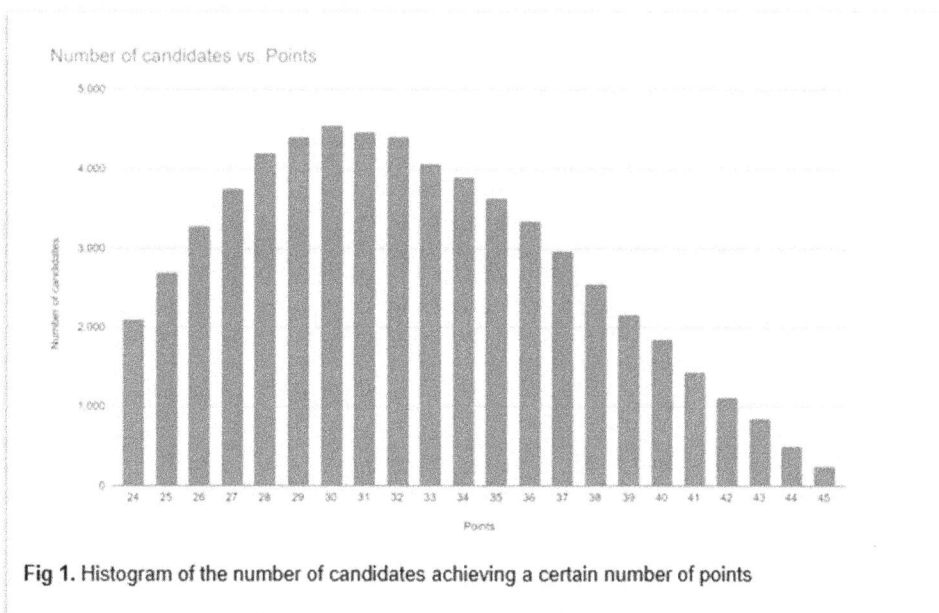

Fig 1. Histogram of the number of candidates achieving a certain number of points

✓ Keep in mind the significant figures as they should remain the same throughout the IA to maintain flow. The estimation should be quoted and done on the required spaces.

> Since I have collected the data from IB website which contains three significant figures, therefore, I will be keeping all my final answers to three significant figures. This would also allow me to keep with the practice for my final exams where I must leave my final answers to three significant figures.

> Since I have used Vernier Calipers to take my measurements, therefore I will keep all my findings and numbers to hundredth part of the mm as this is the least count of the device, I used to collect my data.

Personal Engagement C:

0 The exploration does not reach the standard described by the descriptors below.

1 There is evidence of some personal engagement.

2 There is evidence of significant personal engagement.

3 There is evidence of outstanding personal engagement.

Personal engagement is one of the trickiest criterions as most of the students struggle to execute this criterion on merit. These are some points that can help the student attain maximum score in criterion C.

- ✓ Affiliation with the topic and expectation of the results should be evident throughout the document
- ✓ Shouldn't be very generic
- ✓ An anecdote should complement the topic selection.

35

✓ Collection of data could be secondary or could be primary however collection method should be clear and straightforward.

> **Data Collection and Results**
>
> Using the statistical bulletins provided by the IB, I took the data from the last 5 May examination series from 2015-2019. The reason being that taking into account November series leads to data being skewed since the change in number of candidates is drastic which gives slightly different results. I have taken the averages of the number of candidates and then calculated the percentage and the cumulative frequency. The boundaries of each data point are the number of points the students have achieved while obtaining a diploma, i.e. 24 to 45.

✓ Opinion on the topic should be given at times as the flow of the document shouldn't seem mechanical. You do have the write to wonder or agree to disagree at times. (The document should seem realistic as in everything can't be perfect)

✓ Open-mindedness should surface the document (Perspective's should be welcomed)

✓ The topic shouldn't be very generic (Prefer something unique)

✓ Consistently reinforce why are you glad and excited to work on a certain topic.

Reflection D:

0 The exploration does not reach the standard described by the descriptors below.

1 There is evidence of limited reflection.

2 There is evidence of meaningful reflection.

3 There is substantial evidence of critical reflection.

Reflection is one of the most meticulous criterions where students are required to go deep and evaluate the process on merit. Keeping in mind the following points can bear maximum points in this criterion.

✓ Reflection must occur consistently at each point as it is one of the most common mistakes most student make, as they just restrict it to the evaluation part

✓ Comment on whatever the findings are

✓ Contextualize each result (as in what does r value imply, what does an intercept indicate etc.)

✓ Be honest and discuss the limitations of the process before examiner identifies it

✓ Evaluate the process in term of difficulty and how the experience was.

✓ Mention strengths and weaknesses

✓ Add various approaches and possible extensions.

Use of Mathematics E:

0 The exploration does not reach the standard described by the descriptors below.

1 Some relevant mathematics is used.

2 Some relevant mathematics is used. Limited understanding is demonstrated.

3 Relevant mathematics commensurate with the level of the course is used. Limited understanding is demonstrated

4 Relevant mathematics commensurate with the level of the course is used. The mathematics explored is partially correct. Some knowledge and understanding are demonstrated.

5 Relevant mathematics commensurate with the level of the course is used. The mathematics explored is mostly correct. Good knowledge and understanding are demonstrated.

6 Relevant mathematics commensurate with the level of the course is used. The mathematics explored is correct. Thorough knowledge and understanding are demonstrated.

Criterion E is a make or break for an IA as this determines the caliber and the rigor that has been depicted throughout the process. Following pint can prove to be a grade booster for the IA.

- ✓ Relevance is the key (Randomly discussing a certain concept is detrimental to the fluidity of the document)
- ✓ Clear knowledge and understanding of the applied concept should be demonstrated.
- ✓ The level should be kept on view as in an HL student cannot use a simple mathematical process such as 'finding correlation'. At least two to three concepts should be demonstrated
- ✓ Prefer showing working for each step
- ✓ Be precise and logical
- ✓ Avoid using simple math or just prior knowledge.
- ✓ Clear arguments should be made.

Radial Cycle for the Criterion

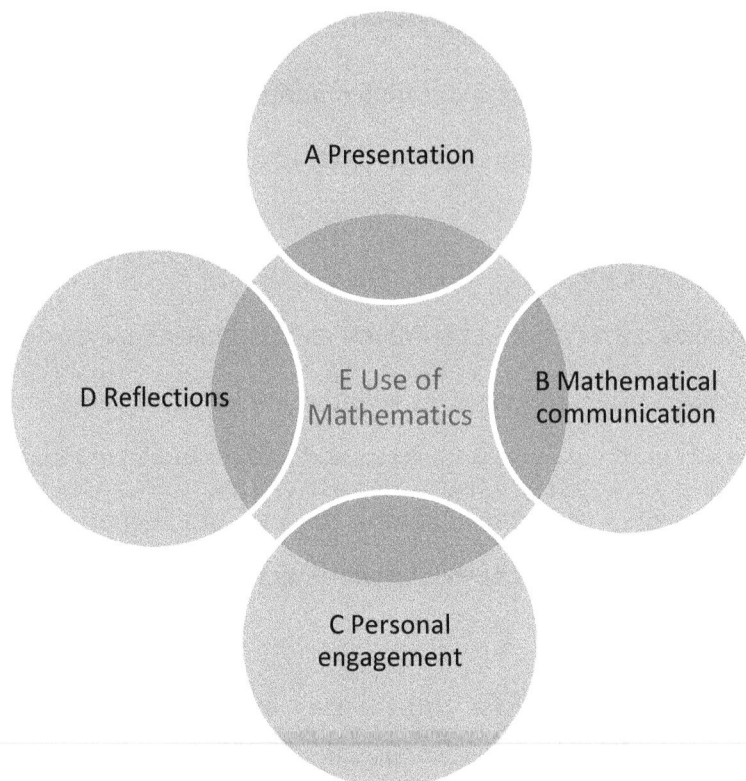

When I was new to IBDP, I used to think that one can get high score in the IA if one scores high in the first four criterions which are general and contain 14 scores out of 20 in total. And then getting 3 out of 6 in criterion A will allow them to score low high. However, later I discovered that criterion E is the core of all the other criterions. I mean how can one score high in presentation, communication, personal engagement, and reflections if the math applied in executing the exploration does not have rigor in it, or it may not be sufficient to meet the level of criterion D. If we do not use rigorous math, we won't get enough of the opportunities to make use of symbols, equations, and mathematical jargons. We can't go deeper into the exploration. We have not much to reflect on to get high scores in the first four criterions. So, while the process of selection of topic, make sure the topic has enough of the depth in it which would allow us to apply appropriate and rigorous math procedure in it to get nothing less than 5 in the use of math, only then we would be able to score good in the other four criterions.

Generally, math IAs can be done in the following areas but not limited to this.

Mathematical modelling

Mathematical modelling is the vast area and it usually considered as the high scoring IAs as it may incorporates all the five topics (number and algebra, functions, geometry and trigonometry, probability and statistics and calculus in it. It can be further divided as.

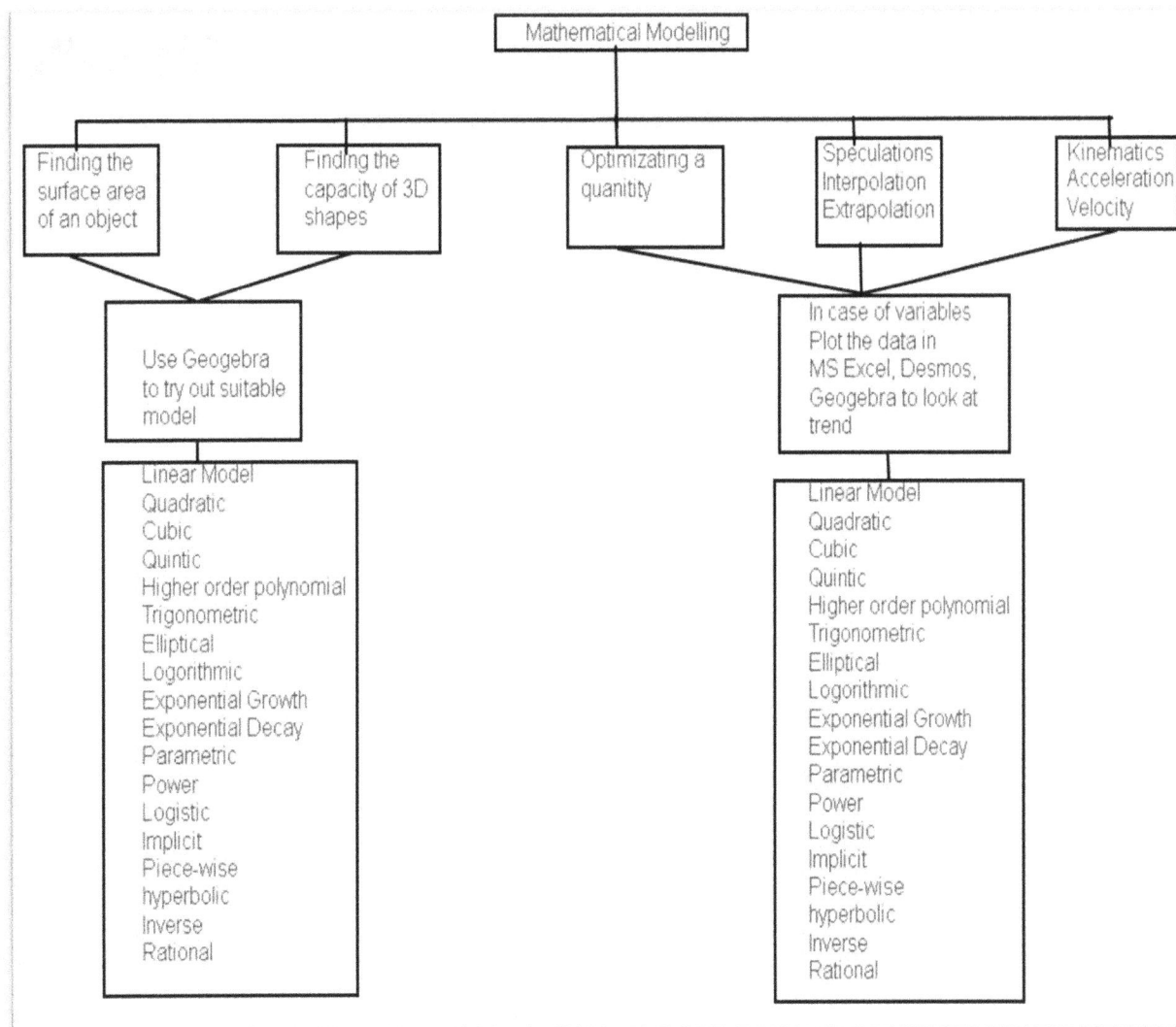

Mathematical Modelling

Finding the surface area of an object

Finding the capacity of 3D shapes

Optimizating a quanitity

Speculations Interpolation Extrapolation

Kinematics Acceleration Velocity

Use Geogebra to try out suitable model

In case of variables Plot the data in MS Excel, Desmos, Geogebra to look at trend

Linear Model
Quadratic
Cubic
Quintic
Higher order polynomial
Trigonometric
Elliptical
Logorithmic
Exponential Growth
Exponential Decay
Parametric
Power
Logistic
Implicit
Piece-wise
hyperbolic
Inverse
Rational

Linear Model
Quadratic
Cubic
Quintic
Higher order polynomial
Trigonometric
Elliptical
Logorithmic
Exponential Growth
Exponential Decay
Parametric
Power
Logistic
Implicit
Piece-wise
hyperbolic
Inverse
Rational

Finding the Surface Area of a Building

A student may think of modelling someone's face, picture, building using GeoGebra, Desmos or any suitable tool and find the best model that fits the situation. AA students tends to use visuals and derive models using standard forms and show all the working stepwise to demonstrate their understanding and use of math. However, AI students can also derive models however, their focus should be on interpretation and analysis of that model. Here is an example of GEMS WORLD ACADEMY where the aim of the exploration is finding the surface area of the front of the building using GeoGebra and then using integration to workout area under the curve.

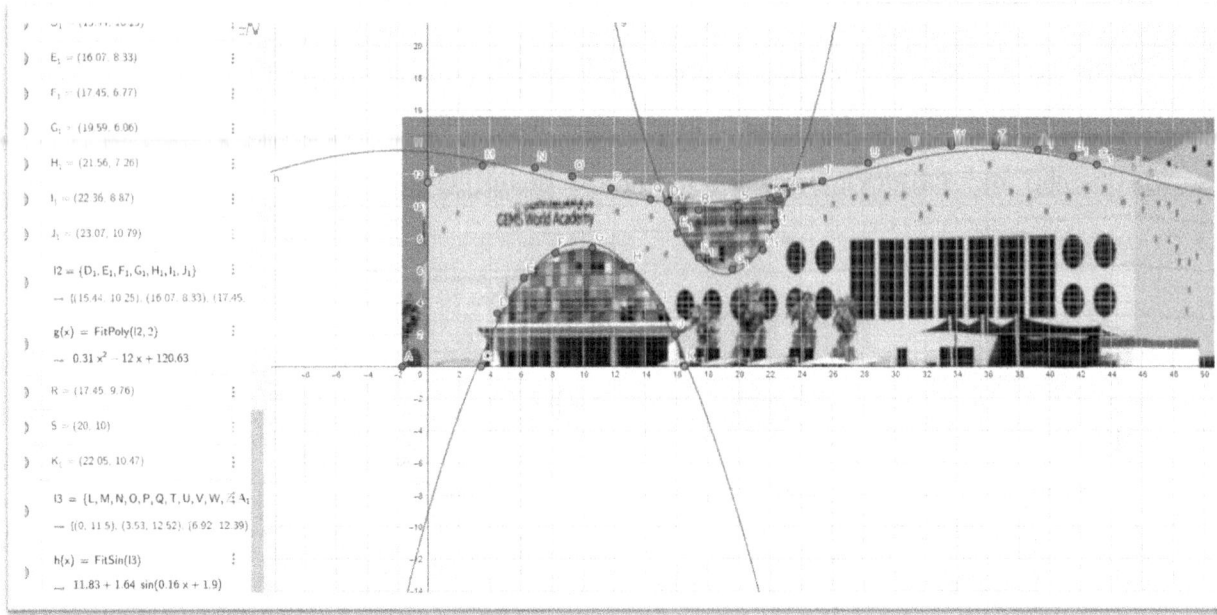

Area under and above the two quadratics can be found using https://www.symbolab.com/. Areas of circular and rectangular windows can be worked out using their formulas. Subtracting this area from sum of all these areas under the sine model will be the required surface area of the front of the building.

Area under the curve for parabola facing down

$$\int_{15}^{23.7} \left(0.31x^2 - 12x + 120.63\right) dx = 56.16981$$

Area under the curve for parabola facing upward

$$\int_{3.2}^{16.8} \left(-0.17x^2 + 3.39x - 9.1\right) dx = 70.4443722$$

Area under the sine curve

$$\int_{0}^{44} \left(1.64\sin(0.16x + 1.9) + 11.83\right) dx = 526.27526\ldots$$

Find the Volume of Revolution of 3D Shapes

Finding the volume of 3D objects can also be done using GeoGebra. Here is an example of finding the volume of a vase. The first step is to model the object. Try out best model that fits the situation. I tried sinusoidal first to see if this fits the best. However, you can see the model looks a little off between points D to G and then from P to Q. We can try another.

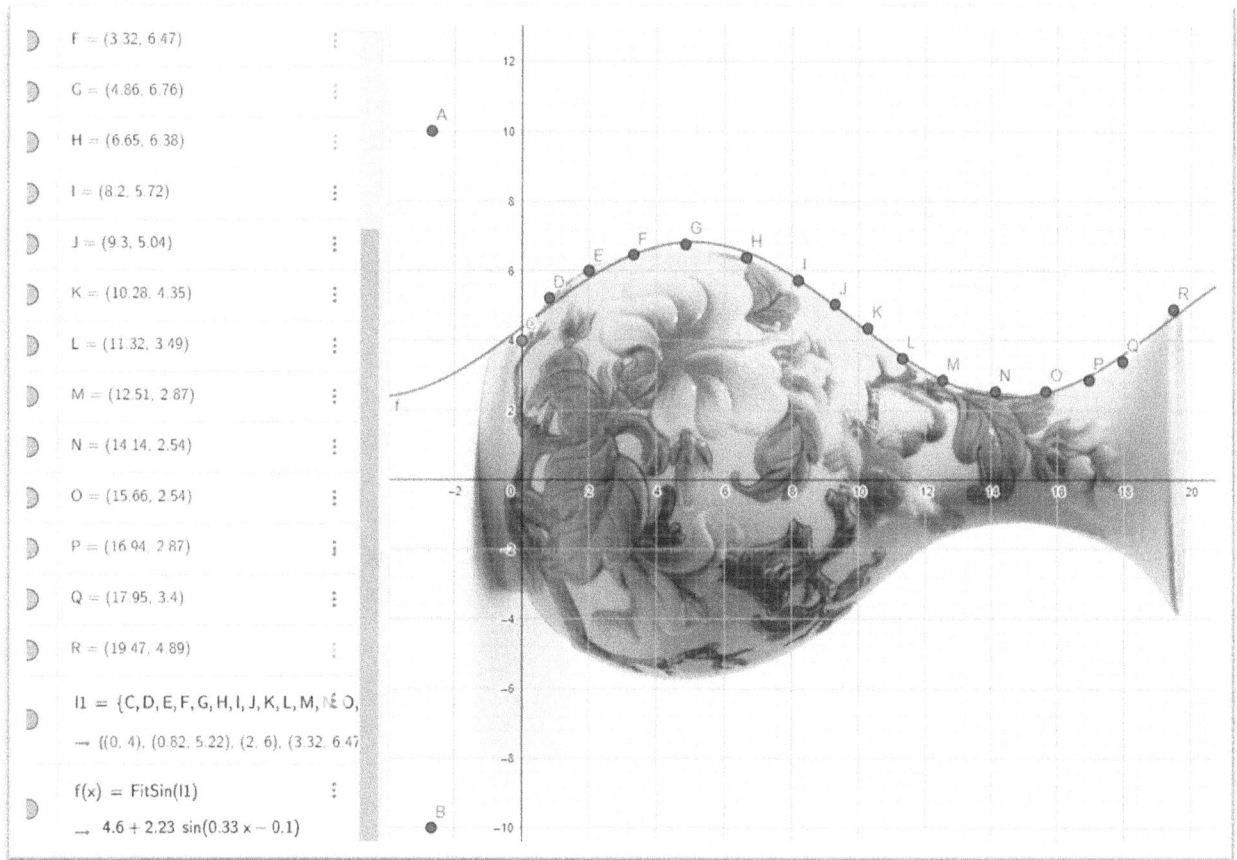

F = (3.32, 6.47)

G = (4.86, 6.76)

H = (6.65, 6.38)

I = (8.2, 5.72)

J = (9.3, 5.04)

K = (10.28, 4.35)

L = (11.32, 3.49)

M = (12.51, 2.87)

N = (14.14, 2.54)

O = (15.66, 2.54)

P = (16.94, 2.87)

Q = (17.95, 3.4)

R = (19.47, 4.89)

l1 = {C, D, E, F, G, H, I, J, K, L, M, N, O,

→ {(0, 4), (0.82, 5.22), (2, 6), (3.32, 6.47

f(x) = FitSin(l1)

→ $4.6 + 2.23 \sin(0.33 x - 0.1)$

Sometimes the model looks very off, and we can reject that model without applying any statistical tool. However, sometimes, the model looks very close in which case we may look for coefficient of determination. The coefficient of determination is **a measurement used to explain how much variability of one factor can be caused by its relationship to another related factor**. This correlation, known as the "goodness of fit," is represented as a value between 0.0 and 1.0. The closer the value is to 1, the better the model is and vice versa. If we look at the trend, cubic function can be a good fit here.

Given below is the cubic model which looks better than sine model visually, however, we will find the coefficient of determination to see which model is best for this vase.

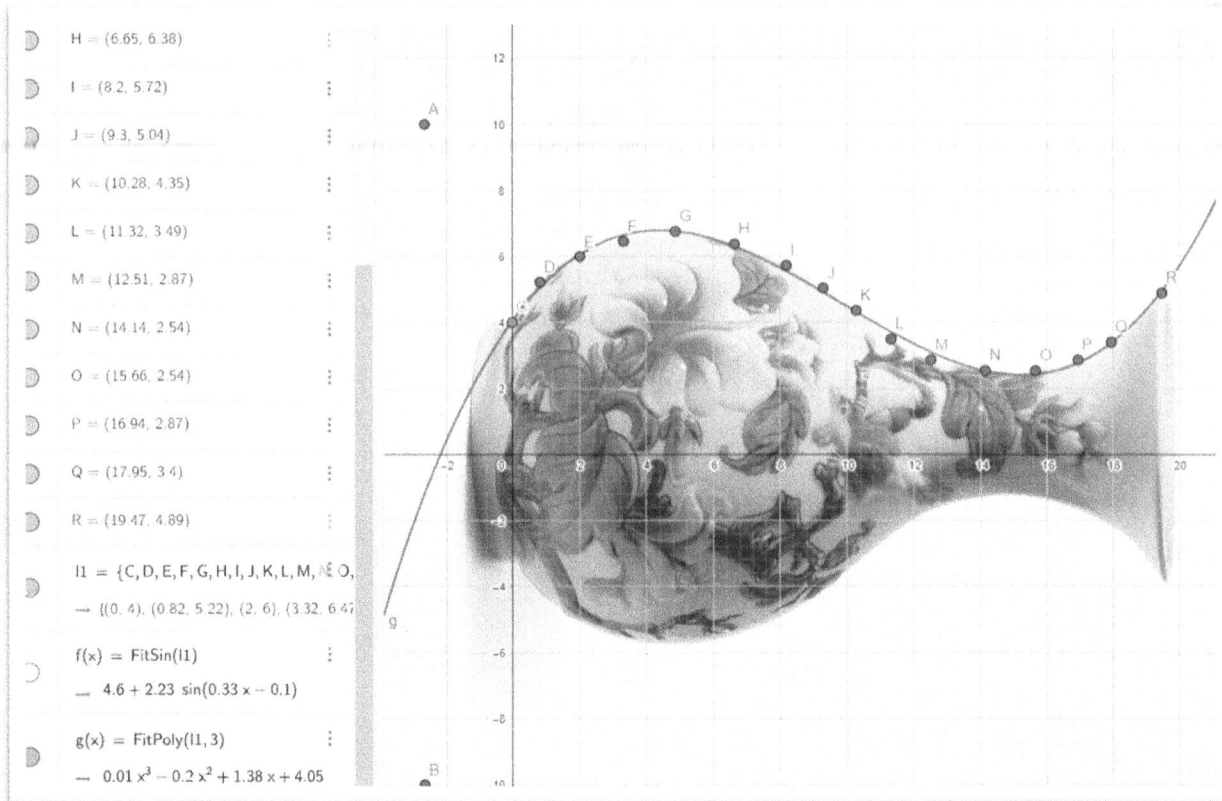

H = (6.65, 6.38)

I = (8.2, 5.72)

J = (9.3, 5.04)

K = (10.28, 4.35)

L = (11.32, 3.49)

M = (12.51, 2.87)

N = (14.14, 2.54)

O = (15.66, 2.54)

P = (16.94, 2.87)

Q = (17.95, 3.4)

R = (19.47, 4.89)

l1 = {C, D, E, F, G, H, I, J, K, L, M, N, O,

→ {(0, 4), (0.82, 5.22), (2, 6), (3.32, 6.47)

f(x) = FitSin(l1)

→ 4.6 + 2.23 sin(0.33 x − 0.1)

g(x) = FitPoly(l1, 3)

→ 0.01 x³ − 0.2 x² + 1.38 x + 4.05

Volume of revolution can be found by putting functions in https://www.symbolab.com/. Since we found that cubic function is best to $\pi \cdot \int_0^{44} \left(0.01x^3 - 0.2x^2 + 1.38x + 4.05\right)^2 dx = 4811397.83624...$ model this, this is how we can find the volume of this vase.

Volume of vase =

Symbolab allows us to show the whole process step-by-step which is important for AA students to show and demonstrate how to get to final answer. Below is the 3D image of vase that we wanted to calculate volume for,

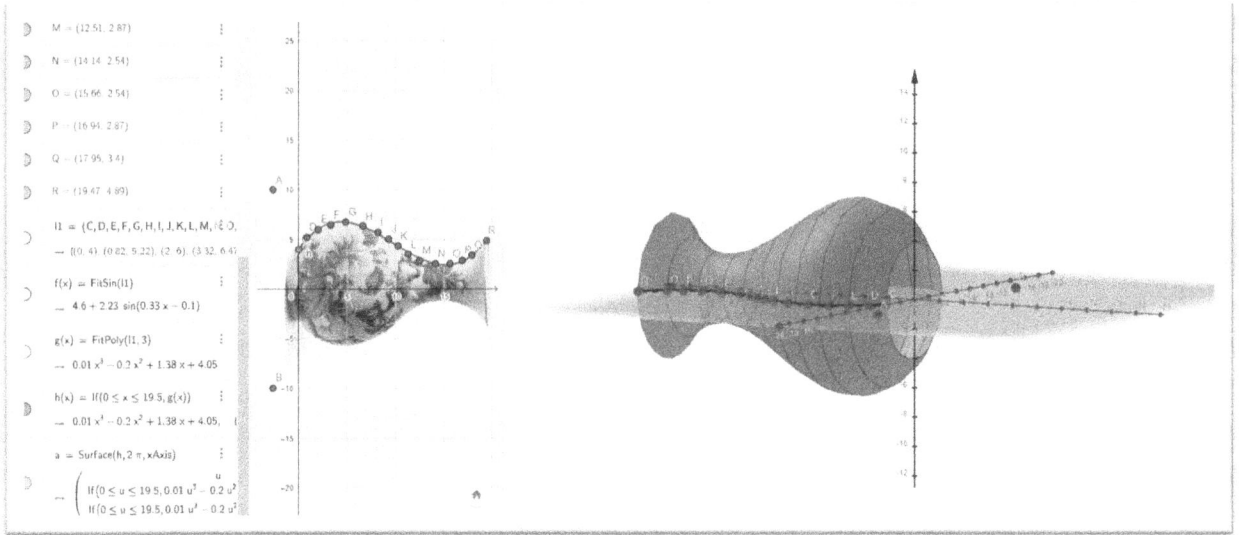

M = (12.51, 2.87)

N = (14.14, 2.54)

O = (15.66, 2.54)

P = (16.94, 2.87)

Q = (17.95, 3.4)

R = (19.47, 4.89)

l1 = {C,D,E,F,G,H,I,J,K,L,M,…}

→ {(0, 4), (0.82, 5.22), (2, 6), (3.32, 6.47),…

f(x) = FitSin(l1)

→ 4.6 + 2.23 sin(0.33 x − 0.1)

g(x) = FitPoly(l1, 3)

→ 0.01 x³ − 0.2 x² + 1.38 x + 4.05

h(x) = If(0 ≤ x ≤ 19.5, g(x))

→ 0.01 x³ − 0.2 x² + 1.38 x + 4.05,

a = Surface(h, 2 π, xAxis)

→ If(0 ≤ u ≤ 19.5, 0.01 u³ − 0.2 u²

    If(0 ≤ u ≤ 19.5, 0.01 u³ − 0.2 u²

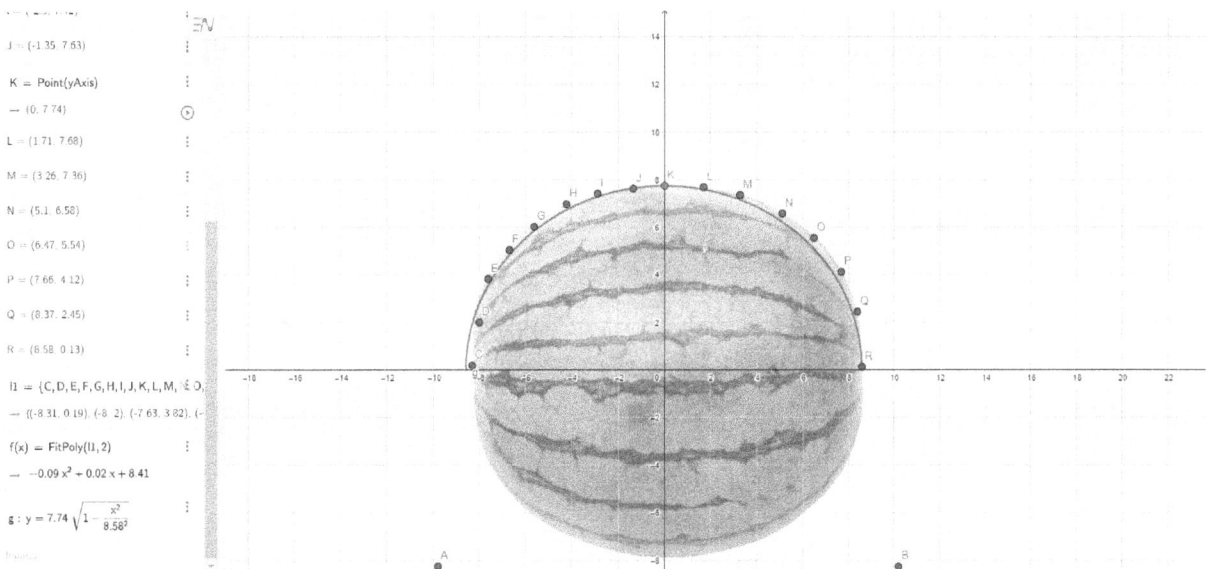

J = (−1.35, 7.63)

K = Point(yAxis)

→ (0, 7.74)

L = (1.71, 7.68)

M = (3.26, 7.36)

N = (5.1, 6.58)

O = (6.47, 5.54)

P = (7.65, 4.12)

Q = (8.37, 2.45)

R = (8.58, 0.13)

l1 = {C,D,E,F,G,H,I,J,K,L,M,…}

→ {(−8.31, 0.19), (−8, 2), (−7.63, 3.82), (−…

f(x) = FitPoly(l1, 2)

→ −0.09 x² + 0.02 x + 8.41

g : y = 7.74 $\sqrt{1 - \dfrac{x^2}{8.58^2}}$

Correlation

Correlation IAs are generally recommended for SL students however, if we use non-linear/transcendental functions in modelling the two variables, this may work for HL too. In order to look for the linear correlation between the two variables, we use Pearson's Product Moment correlation commonly known as 'r' value ranging from -1 to +1.

Facts about r

- The value of r is always in the interval $-1 \leqslant r \leqslant +1$
- The sign of r tells us the direction of the correlation: positive or negative or zero.
- The size of r tells us the strength of the linear correlation, as shown.

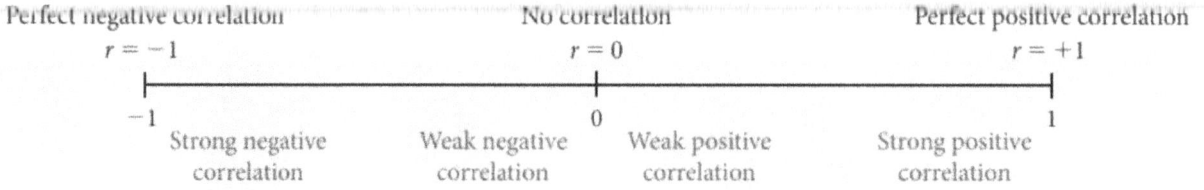

Perfect negative correlation	No correlation	Perfect positive correlation
$r = -1$	$r = 0$	$r = +1$

$$-1 \qquad\qquad 0 \qquad\qquad 1$$

| Strong negative correlation | Weak negative correlation | Weak positive correlation | Strong positive correlation |

- If $r = +1$, there is a perfect positive linear correlation; all the points fall on a line with positive slope.
- If $r = 0$, there is zero correlation.
- If $r = -1$, there is a perfect negative linear correlation; all the points fall on a line with negative slope.
- r has no units, and is not a percentage.

If the correlation is not linear, we use Spearman's rank correlation coefficient, r_s,

Facts about r_s
- The value of r_s is always in the interval $-1 \leqslant r \leqslant +1$
- The sign of r_s tells us the direction of the rank-order correlation: positive or negative or zero.
- The size of r_s tells us the strength of the rank-order correlation.
- If $r_s = +1$, there is a perfect positive rank-order correlation.
- If $r = 0$, there is zero correlation.
- If $r = -1$, there is a perfect negative rank-order correlation.
- r has no units, and is not a percentage.
- r is a measure of monotonicity.

Data Analysis

Voronoi Diagrams

Voronoi diagrams can be used to find the best location to open a store. For example, if we want to open a coffee shop in London, we can spot all the coffee shops in vicinity and can work out the place that is at a maximum possible distance from all the existing coffee shops in the town.

Probability and Distributions

In statistics, IAs can be picked from normal distribution, binomial distribution, poison distribution and probability density functions.

Phenomenon which are normally distributed are

- Height
- Rolling some dice
- Tossing a coin
- IQ level of students
- SAT score
- IB Score
- Stock Market
- Income distribution in economy
- Size of the shoes
- Birth Weight
- Weight of canned juice
- Bag of cookies

And many more phenomenon which can lead to normal distribution. Phenomenon which can be modelled using binomial distribution are as follows

- Number of people responding 'Yes' to a particular survey.

- Number of patients responding to a specific medicine
- Flipping a coin for n times where n is greater or 50.
- Number of games win by a team in a tournament.
- Number of votes won by a candidate in an election.
- Number of defective products in a production run.

An exemplar.

My first objective is to produce a probability distribution function, to determine the probability of a certain coach having a certain number of years. I started off by producing a table using the data collected above. The table shows the number of coaches in my sample that were head coach within a certain interval of time (in years). For example, the number of coaches from the data collected above that spent less than a year coaching was 17. The length of each interval was kept equal, and the length of

Using the obtained probability for each interval, we can produce a graph representing the plots of the probability distribution function that we will define later on in this exploration:

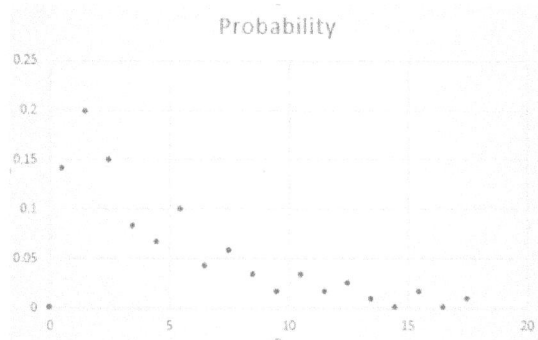

Probability

Two different functions were made through the use of technology (excel), and a PDF was produced as a result. The PDF is:

$$f(x) = \begin{cases} 0 & x \le 0 \\ 0.0003x^5 - 0.0066x^4 + 0.0588x^3 - 0.2404x^2 + 0.3828x + 0.0006 & 0 < x \le 5.379 \\ 0.0008x^2 - 0.0238x + 0.1859 & 5.379 < x \le 18 \\ 0 & x > 18 \end{cases}$$

And the following is the graph, made on Desmos, of the PDF:

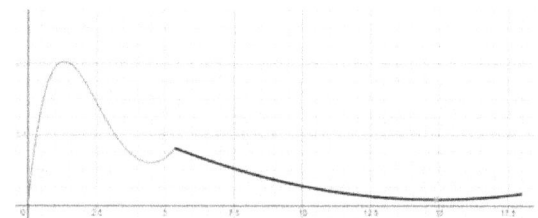

each interval was chosen by considering what number of intervals would allow us to produce a somewhat compact and still somewhat precise table. 1 year seemed to be the most reasonable interval length. The frequency of the data in this case is the number of coaches that fall under a certain time interval.

Frequently Asked Questions and Answers.

Selection of the topic

Q1. What is the difference between math EE and math IA?

EE and IA are almost similar in terms of what they expect from student however, EE as name suggests is an extended and much more in-depth and more complex analysis of your topic, whereas, IA is much more focused, straight forward and have limited possibilities in exploration. The difference also lies in the word limit and the criterions as well.

Q2. Does the selection of topic vary for AA/AI math?

There is no such difference in terms of the selection of the topic for the math AA and AI. The difference lies in the approach and execution of the exploration. AI is more technology driven and focuses more on the application and interpretation of mathematical techniques. Whereas math AA is more of demonstrating on how formulas, equations and models are derived followed by the interpretation in the specific context. AA requires students to show all the mathematical steps involved in the derivation of their mathematical procedures.

Q3. At what level the selection of topic varies in terms of writing an IA for SL and HL?

While picking up a topic for the HL, one should make sure that the topic provides enough of the depth which allows one to apply rigorous (At least one math procedure from HL). We can apply math on almost everything but the question one should ask oneself is whether the topic meets the rigor of criterion E. In case of SL, make sure the topic should not limit to prior knowledge math.

Q4. Can two students use same topic for exploration in one class?

Yes, they can, however, the exploration must be different in terms of data, approach, and findings. They can work on two different avenues within the same topic.

Presentation

Q1. What should be on the first page? Is table of content an important component for the IA?

First page should have a topic and a specific title of that topic, for example, putting just 'Modeling' as a topic is not sufficient. It is better to write' Can a model made from the previous years be used to determine the percentage or number of candidates that will achieve a certain grade? Table of content is not mandatory. Put no of pages.

Q2. Can we submit handwritten IA?

Students can write text, equations, and draw graphs and table by hand, however, word processed is encouraged and looks more professional. And there are lot of tools available now-a-days which help draw graphs, use symbols at ease.

Q3. Can student use external resources such as MS office, Desmos, GeoGebra etc.?

Yes of course, students can use external sources and should acknowledge these in their IA as the skills they learnt while writing the IA.

Q4. Do we need any cover page as we do in EE?

No page cover is required in the IA.

Q5. Does an exploration have to be less than 20 pages long to be concise?

Not necessarily, it depends on the topic of the IA and the approach. If justified, relevant and necessary math procedures, calculations, graphs, or interpretation are used in the IA, no of pages can go little high.

Q6. Is bibliography or appendices included in the page count?

No, it does not.

Mathematical Communication

Q1. What should be the target audience for the student?

The target audience for this exploration should be fellow students. Student should request their fellow members to read their IAs followed by a small viva to double check if they understand the IA in similar context.

Q2. What do we mean by defining the key terms in math IA?

It means defining every term that is not commonly known, for example, define what is income per capita, GDP, scoring an Ace in tennis, body mass index etc. Don't add anything unexplained that a student must google it to know.

Q3. If the IA does not contain graph, charts, or tables, can we still earn some score in criterion B?

It is recommended to use multiple forms of mathematical procedures but not necessary to use all forms. Wherever possible, use all appropriate ways of showing information, models, equations, and formulas.

Personal Engagement

Q1. Does personal engagement mean writing a good and convincing start why to pick a specific topic?

There is common misconception that personal engagement means student need to have a personal vested interest in the topic which leads them writing deliberate lies which can reveal on examiner at ease. Personal engagement demands the topic to be unique and independent. It should have unique approach of exploring the topic. It should display some degree of creativity by exploring the varied aspects of the IA and discuss unique avenues.

Reflections

Q1. What constitute reflections in the IA?

Discussing the challenges faced by the students and how those overcome using various skills mentioned in the learner profile. Reflections can be done at various level:

- Reflection on how to collect the data, and if there is any specific sampling technique used and why.
- Doing commentary on your results and equations
- Discussing the limitations and assumptions, interpolation, and extrapolations.
- What went good and what bad, critically assessing it.
- Writing possible extensions of the IA.

Q2. What is the difference between a conclusion and a reflection?

A conclusion could be more descriptive and factual. It is wrapping up of what is said in the aim and rationale of the IA. Conclusion comes in the end of the IA.

Whereas reflection is an ongoing process in the exploration and must be seen throughout the work. It runs throughout the IA.

Use of Mathematics

Q1. Should students use math that is beyond IB course?

IB does not encourage to use math beyond the IB course. If there is a need to use math outside the course, it is advisable that the level of the math used should be commensurate with the level of the IB.

Q2. What is the difference in criterion E (use of the math) in SL/HL IA?

HL students are expected to use at least one math procedure that should be purely from IB math.

Q3. How can we assure if it is a math? (Not a physics or economic IA)

Make sure you extend math in your IA. Your IA should be math heavy.

Q4. Can I derive the already derived equations like Reimann's sum?

IB expect students to apply already discovered math on real life scenarios. Proofs like Reimann's sum is already done however, if you have a different way of proving it, you can take this as your IA.

PART II

SEVEN EXAMPLES OF EXCELLENT INTERNAL ASSESSMENTS

The assessments featured in this section are all recently submitted IA that scored exceptionally after being moderated by the IBO. To prevent plagiarism and duplication of results, the appendices have been omitted. The IAs are presented in the exact same way as they were submitted, and without any edits or changes to formatting. We do not retain the copyright of these commentaries, nor is this publication endorsed by the IBO. The Internal Assessments are being re-printed with the permission of the original authors.

1. MODELLING THE DEPRECIATION OF COCOA BEANS IN MALAYSIA

Author: Caitlyn Tan
Moderated Mark: 18/20
Level: Math AI HL

Introduction

Chocolate is harvested from the cocoa bean and in recent years, cocoa bean production globally has skyrocketed. While it is often grown in West African regions, other countries with suitable climates have also begun increasing their own cocoa production. An example of this would be Malaysia, whose warm tropical climate is suitable for the growth of Malvaceae (cocoa trees). Currently the 33rd highest chocolate producer in the world (worldatlas.com, 2015), Malaysia's ideal agricultural climate poses significant implications for the future of the cocoa industry there. However, it is currently being predicted that the cocoa production is to rapidly decline overtime. This is due to the fact that "cocoa can only grow in regions with intense heat year-round and rainforest-like conditions" (Austin, 2018), and climate change could pose an issue to cocoa producing countries.

Not only is chocolate widely consumed globally with 7.7 million metric tonnes consumed worldwide in 2019 (Statista, n.d., 2017), it is also an ingredient that is particularly close to my heart. As an avid baker, I find myself using chocolate often and in a variety of forms, from chips to bars. Singapore does not have the space or agricultural capabilities to cultivate cocoa trees, therefore all of our chocolate is imported. Choosing to buy chocolate from Malaysia would be cheaper and more environmentally conscious due to the short distance between Singapore and Malaysia, leading to lower import costs and less production of fossil fuels during transport. Online data has shown that from 1990, cocoa production in Malaysia has been on the decline, which would have an impact on the availability of chocolate in Singapore. Therefore, the future of cocoa production in Malaysia is especially relevant to me as it would directly impact my biggest hobby which is also shared by many Singaporeans.

This exploration will aim to model the declining production of cocoa beans in Malaysia until 2050, which is when the Intergovernmental Panel on Climate Change (IPCC) has predicted cocoa to go extinct by (Austin, 2018). This is due to climate change and other ecological conditions such as pests and plant diseases, and as these problems worsen, suitable cocoa cultivation areas will drop overtime (Vishnefske, 2018).

Methodology
Choice of sampling

In order to determine the future of cocoa production in Malaysia, records of this production must first be obtained. I have chosen to use the online database made available by ourworldindata.org. It is a trusted website which is used by reputable news sources such as The BBC, The New York Times and The Washington Post, as well as colleges such as Harvard University and the University of Cambridge. The raw data will be graphed and then relevant data points will be selected to construct a model in order to predict the future of the cocoa industry in Malaysia.

Mathematics used

The exploration will include relevant data selected from the sample. A scatter plot will be created using this data, the overall trends will be identified and the nature of the function will be determined.. From this, 2 coordinates will be randomly selected in order to develop an initial model. This will be done by substituting the x and y values of these coordinates into the equation to obtain a model. Seeing as this will be the first attempt, the model may not fit the data perfectly, but will act as a good starting point.

In order to try to obtain a more accurate model, several coordinates will be used to find many a and b values. The mean average of these values will be used to form a more accurate equation. Following this, the observed and expected values will be compared using Pearson's Correlation Coefficient, the expected values having been calculated from the new model, in order to determine the true accuracy of the model obtained from the average a and b values. The closer the r-value is to 1, the more accurate the function is in modelling the future of cocoa production in Malaysia due to the strong positive correlation between the two values.

Key definitions

➢ **Cocoa** - The cocoa in this exploration refers to the cocoa bean, or the malvaceae plant, in which cocoa solids, cocoa powder, cocoa nibs and other similar products are derived from in order to make chocolate or chocolate-like products.

➢ **Exponential function** - A function in the form $f(x) = a \times b^x + c$, where the variable x occurs as the exponent of a, and b is a positive real number (britannica.com).

➢ **Growth and decay** - A situation in which quantities are either increasing (growth) or decreasing (decay) exponentially.

➢ **Pearson's correlation coefficient** - The statistical test that measures the statistical relationship or association between two continuous variables (statisticssolutions.com, 2019).
  ○ For a set of n data given as ordered pairs (x_1, y_1) (x_2, y_2), (x_3, y_3), ...
  ○ Pearson's correlation coefficient is $r = \frac{\Sigma(x-\bar{x})(y-\bar{y})}{\sqrt{\Sigma(x-\bar{x})^2 \Sigma(y-\bar{y})^2}}$, where $\bar{x}$ and $\bar{y}$ are the means of the x and y data respectively and Σ means the sum over all the data values.

Presentation of data
Graphed raw data

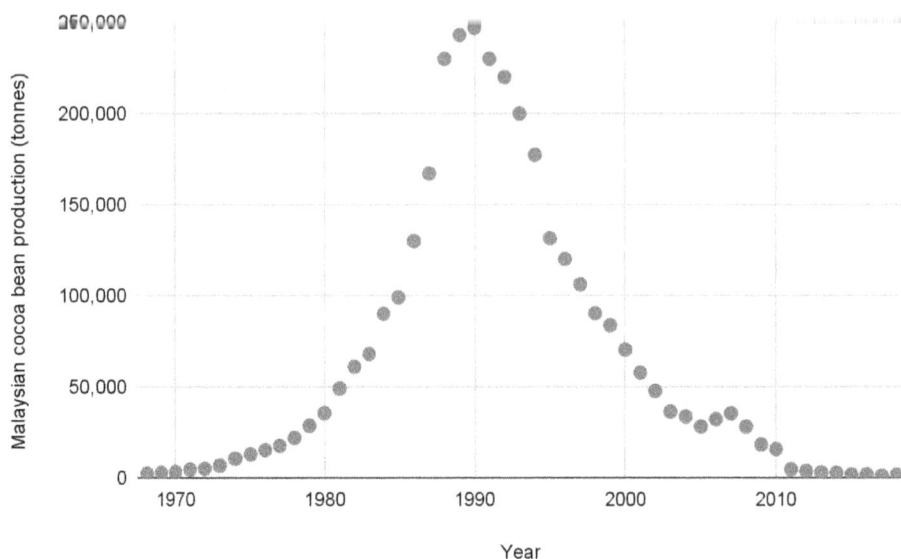

Graph 1: Scatter plot showing the Malaysian cocoa bean production in tonnes over a 50 year period, from 1968 to 2018. The full table with data from 1968 to 2018 can be found in the appendix.

From graph 1, it can be seen that between the years of 1968 and 1990, there was a sharp increase in the tonnes of cocoa beans produced in Malaysia per year. This curve seems to resemble that of an exponential function demonstrating exponential growth. However, a steep and nearly continuous decline in production followed this, with only a brief increase between the years 2005 and 2007. This could have been due to a variety of reasons, such as climate change altering the agricultural conditions cocoa is grown in through "long periods of drought... decreas[ing] soil fertility" (Hutchins et al., 2015). Another reason could be an alteration in the regulations which allow a product to be classified as chocolate, thereby reducing the demand of cocoa as a product (Santa Barbara Chocolate, 2017).

For the purposes of this exploration, cocoa production in Malaysia will only be looked at following 1990, the beginning of the decline. This is because the data demonstrates that the cocoa production will likely continue its decline in the years following this, and I aim to calculate a function which will model this future decline. However, I have chosen to include the brief increase in cocoa production between the years 2005 and 2007 because despite not following the general trend, they could represent a potential for a future increase. An exponential function will be used as a model because the data seems to follow the trend of exponential decay.

Data relevant to the exploration

x (Years from 1990)	y (Malaysian cocoa bean production in tonnes)
0	247000
1	230000
2	220000
3	200000
4	177172
5	131475
6	120071
7	106027
8	90183
9	83668
10	70262
11	57708
12	47661
13	36236
14	33423
15	27964
16	31937
17	35180
18	27955
19	18152
20	15654
21	4605
22	3645
23	2809
24	2665
25	1729
26	1757
27	1029
28	1505

Table 1: Table showing the relevant data from 1990 to 2018 rounded to the nearest whole number, whereby cocoa production is decreasing.

The data available on ourworldindata.com is from 1961 to 2018. I chose to select data from 1968 to 2018 in order to specifically obtain a sample size of 50. Neglecting data from the years between 1961 and 1968 is in no way detrimental to the exploration as this data follows the general upwards trend as demonstrated by the data between 1968 and 1990.

Relevant data graphed

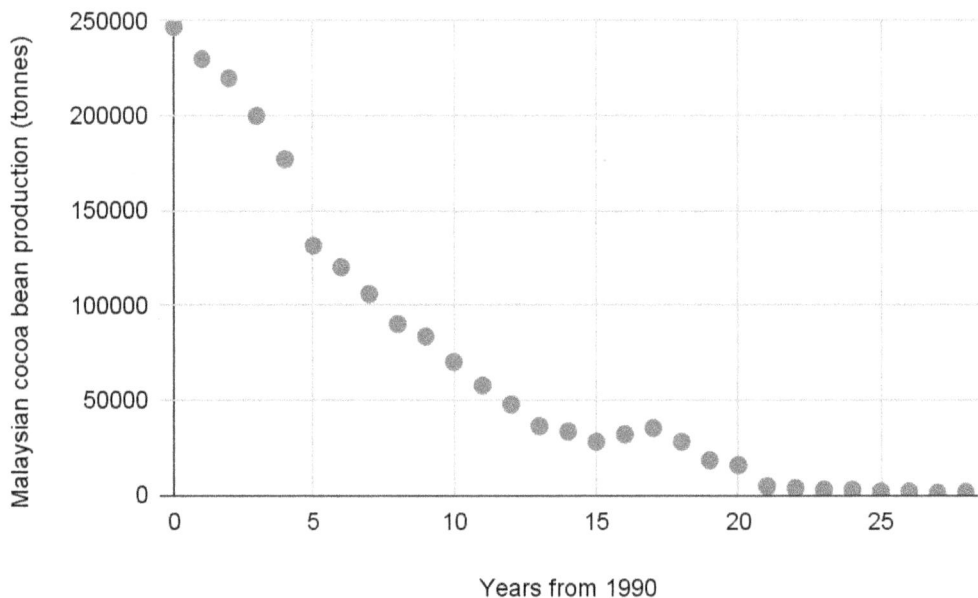

Graph 2: Scatter plot showing Malaysian cocoa bean production in tonnes against the years from 1990.

The points plotted on this graph most closely resemble a decaying exponential function which takes the form $f(x) = a \times b^x + c$, whereby the x variable occurs as an exponent. The notation $f(x) = a \times b^x + c$ will be specifically used for models, while $y = a \times b^x + c$ will be used for calculations. The graph has an asymptote of 0 as cocoa production could never drop it below 0 tonnes per year. Therefore, within the exponential equation, the value of c will always be equal to 0.

Calculations
Finding an initial exponential equation with 2 points
To get an initial exponential equation, I selected 2 random coordinates from my data set using an online random number generator. These points were (0, 247000) and (19, 18152). The cocoa yielded in tonnes will function as the y value, or the dependent variable, and the years from 1990 will function as the x value, or the independent variable. To find the

exponential function, I inputted the first coordinate (0, 247000) into the equation $y = a \times b^x$ to find the value of a in terms of b. Due to the rules of exponents, I was able to find the exact numerical value of a rather than find a in terms of b. I then inputted the second coordinate (19, 18152) into the equation $y = a \times b^x$, substituting the pre-calculated a value into the equation and solving for b. Finally, I used the now known a and b values from the equation to form my initial model in the form $f(x) = a \times b^x$. This model was then plotted against my data points to assess the fit.

$(0, 247000)$	First set of data points to substitute into $y = a \times b^x$
$247000 = a \times b^0$	Substituting the x and y values of the coordinates into $y = a \times b^x$ to calculate a in terms of b
$a = \frac{247000}{b^0}$	$b^0 = 1$ according to the laws of exponents
$a = \frac{247000}{1}$	Anything divided by 1 always equals itself, $\frac{y}{1} = y$
$a = 247000$	Anything divided by 1 always equals itself, $\frac{y}{1} = y$
$(19, 18152)$	Second set of data points to substitute into $y = a \times b^x$
$18152 = a \times b^{19}$	Substituting the x and y values of the coordinates into $y = a \times b^x$
$18152 = 247000 \times b^{19}$	Substituting the a value obtained from the first set of data points
$\frac{18152}{247000} = b^{19}$	If $y = a \times b$, then $b = \frac{y}{a}$
$\sqrt[19]{\frac{18152}{24700}} = b$	If $b^{19} = \frac{y}{a}$, then $\sqrt[19]{\frac{y}{a}} = b$
$b \approx 0.872$	Therefore $b = 0.872$ when rounded to 3 significant figures

By substituting the obtained a and b values from the coordinates into the function $f(x) = a \times b^x$, it can be seen that the initial exponential function would be $f(x) = 247000 \times 0.872^x$. This will be **model 1** (the initial model). 0.872 as a rounded figure will be used as this enables shorter calculations while also maintaining the overall accuracy of the function.

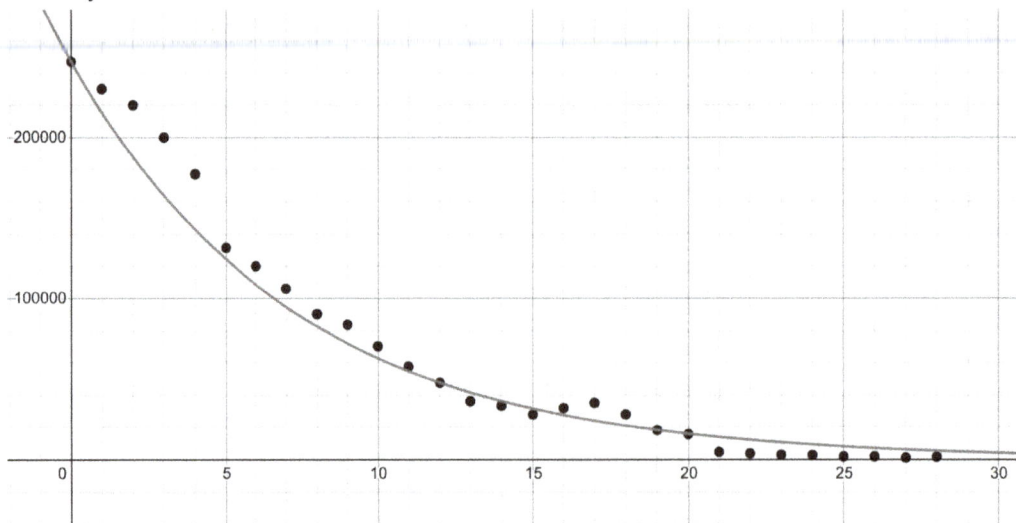

Graph 3: Model 1, $f(x) = 247000 \times 0.872^x$ plotted against the relevant data points. x-axis is years from 1990 and y-axis is cocoa production in tonnes.

Visually, this model does in fact seem to fit the data well, with the coordinates in the data set being very close to the exponential function. It also goes through the points $(0, 247000)$ and $(19, 18152)$ since these coordinates were used to obtain the model. However, in order to determine the true accuracy of the initial equation, the expected values need to be calculated. This will be done through substituting the x values of the coordinates from the raw data table into the initial model. From there, the correlation between the observed values and the expected values will be determined.

Eg. In 1997 (7 years after 1990), 106027 tonnes of cocoa was produced. For the coordinates (7, 106027), the x value 7 will be substituted into the expression 247000×0.872^x to obtain a new y value. This is the expected value of cocoa production for that year, and will be compared against the observed 106027 tonnes.

$y = 247000 \times 0.872^x$

$= 247000 \times 0.872^7$

≈ 94692 (rounded to the nearest whole number)

Therefore, according to model 1, the expected amount of cocoa produced in Malaysia in tonnes in 1997 is 94692 tonnes.

x (years)	O_y (tonnes)	E_y (tonnes)
0	247000	247000
1	230000	215384
2	220000	187815
3	200000	163775
4	177172	142811
5	131475	124532
6	120071	108592
7	106027	94692
8	90183	82571
9	83668	72002
10	70262	62786
11	57708	54749
12	47661	47741
13	36236	41630
14	33423	36302
15	27964	31655
16	31937	27603
17	35180	24070
18	27955	20989
19	18152	18302
20	15654	15960
21	4605	13917
22	3645	12136
23	2809	10582
24	2665	9228
25	1729	8047
26	1757	7017
27	1029	6118
28	1505	5335

Table 2: Table showing the observed (O_y) and expected (E_y) values, rounded to the nearest whole number, obtained from model 1, to be compared using Pearson's correlation coefficient.

The r-value

The following table is a guide for describing the strength of linear correlation using r.

Positive correlation			Negative correlation		
$r = 1$	perfect positive correlation		$r = -1$	perfect negative correlation	
$0.95 \leqslant r < 1$	very strong positive correlation		$-1 < r \leqslant -0.95$	very strong negative correlation	
$0.87 \leqslant r < 0.95$	strong positive correlation		$-0.95 < r \leqslant -0.87$	strong negative correlation	
$0.7 \leqslant r < 0.87$	moderate positive correlation		$-0.87 < r \leqslant -0.7$	moderate negative correlation	
$0.5 \leqslant r < 0.7$	weak positive correlation		$-0.7 < r \leqslant -0.5$	weak negative correlation	
$0 < r < 0.5$	very weak positive correlation		$-0.5 < r < 0$	very weak negative correlation	

Fig. 1: Table illustrating different correlation strengths depending on the r-value via Haese Mathematics.

As previously stated, Pearson's correlation coefficient will be used in order to determine the correlation between the observed and expected values. Correlation is determined by the r-value, which is found using the equation $r = \dfrac{\Sigma(x-\bar{x})(y-\bar{y})}{\sqrt{\Sigma(x-\bar{x})^2\Sigma(y-\bar{y})^2}}$.

The r-value can affect the strength of the correlation. For example, an r-value of 1 would mean there is a perfect positive linear correlation between the two variables. An r-value of 0 would mean that there is no relationship whatsoever between the two variables. In terms of the model, the closer the r-value is to 1, the more accurate the model would be as this would be an indicator that there is a strong positive linear correlation between the observed and expected values.

Within the equation to calculate Pearson's correlation, the Σ symbol represents the sum of the values following it. $\bar{x}$ refers to the mean average of the x values and $\bar{y}$ refers to the average of the y values. Relating the equation back to the initial model, x will be the notation used for the observed values (O_y) while y will be the notation used for the expected values (E_y).

	r=	x	Oy	Ey	Oy-Ō	Ey-Ê	(Oy-Ō)^2	(Ey-Ê)^2	(Oy-Ō)(Ey-Ê)
2	0.9930598938	0	247000	247000	177087.1724	181712.3864	31359866634	33019391364	32178932697
3		1	230000	215384	160087.1724	150096.3864	25627902771	22528925205	24028506085
4		2	220000	187814.848	150087.1724	122527.2344	22526159323	15012923165	18389766152
5		3	200000	163774.5475	130087.1724	98486.93384	16922672427	9699676137	12811886743
6		4	177172	142811.4054	107259.1724	77523.79176	11504530067	6009938289	8315137747
7		5	131475	124531.5455	61562.17241	59243.93187	3789901072	3509843464	3647185148
8		6	120071	108591.5077	50158.17241	43303.89405	2515842260	1875227240	2172044184
9		7	106027	94691.79469	36114.17241	29404.18107	1304233449	864605864.3	1061907665
10		8	90183	82571.24497	20270.17241	17283.63135	410879889.7	298723912.6	350342187.4
11		9	83668	72002.12561	13755.17241	6714.511993	189204768.1	45084671.3	92359270.14
12		10	70262	62785.85353	349.1724138	-2501.760086	121921.3746	6258803.525	-873545.6078
13		11	57708	54749.26428	-12204.82759	-10538.34934	148957816.4	111056806.8	128618736.7
14		12	47661	47741.35845	-22251.82759	-17546.25517	495143830.9	307871070.3	390436244.7
15		13	36236	41630.46457	-33676.82759	-23657.14905	1134128716	559660701.1	796697729.7
16		14	33423	36301.76511	-36489.82759	-28985.84851	1331507517	840179414	1057688615
17		15	27964	31655.13917	-41948.82759	-33632.47445	1759704136	1131143337	1410842872
18		16	31937	27603.28136	-37975.82759	-37684.33226	1442163481	1420108898	1431093705
19		17	35180	24070.06134	-34732.82759	-41217.55227	1206369312	1698886616	1431602137
20		18	27955	20989.09349	-41957.82759	-44298.52013	1760459296	1962358885	1858669670
21		19	18152	18302.48953	-51760.82759	-46985.12409	2679183272	2207601886	2431988907
22		20	15654	15959.77087	-54258.82759	-49327.84275	2944020371	2433236071	2676470915
23		21	4605	13916.9202	-65307.82759	-51370.69342	4265112344	2638948143	3354908389
24		22	3645	12135.55441	-66267.82759	-53152.05921	4391424973	2825141398	3522271496
25		23	2809	10582.20345	-67103.82759	-54705.41017	4502923677	2992681902	3670942412
26		24	2665	9227.681405	-67247.82759	-56059.93221	4522270315	3142716000	3769908656
27		25	1729	8046.538185	-68183.82759	-57241.07543	4649034344	3276540717	3902915618
28		26	1757	7016.581297	-68155.82759	-58271.03232	4645216834	3395513208	3971510432
29		27	1029	6118.458891	-68883.82759	-59169.15473	4744981703	3500988871	4075797853
30		28	1505	5335.296153	-68407.82759	-59952.31747	4679630875	3594280370	4101207797
31	AVERAGE		69913	65287.61362					
32	SUM/SIGMA						167453547396	130909512409	147030766516

Fig. 2: Screenshot of the calculations for model 1 done on google sheets. Cells highlighted in blue are the average of the above values and cells highlighted in green are the sum of the above values. Cell highlighted in red is the final r-value calculated according to the equation $r = \frac{\Sigma(x-x)(y-y)}{\sqrt{\Sigma(x-x)^2\Sigma(y-y)^2}}$.

According to the above calculations, it can be seen that the r-value of model 1 is equal to 0.993 when rounded to the nearest 3 significant figures. This demonstrates that there is a very strong positive correlation between the observed and expected values. As a result, the model $f(x) = 247000 \times 0.872^x$ is a highly accurate exponential function to model the future of cocoa production in Malaysia.

Refining the model

Despite the high r-value of 0.993 obtained from model 1, a new and ideally more accurate model will still be calculated. This is because the 2 points used to calculate model 1 were chosen at random which could hinder the correlation between the observed and expected values extracted from the model.

In order to revise the model, 5 random pairs of coordinates will be selected by an online algorithm. No 2 coordinates will be repeated. From each pair, an a and b value will be calculated, similar to what was done with model 1. Finally, the average of these values will be calculated to use in the final revised model.

$(14, 33423)$	First set of coordinates to substitute into $y = a \times b^x$
$33423 = a \times b^{14}$	Substituting the x and y values into the equation $y = a \times b^x$ to calculate a in terms of b
$a = \frac{33423}{b^{14}}$	If $y = a \times b^x$, then $a = \frac{y}{b^x}$
$(3, 200000)$	Second set of coordinates to substitute into $y = a \times b^x$
$200000 = a \times b^3$	Substituting the x and y values into the equation $y = a \times b^x$ to calculate a in terms of b
$200000 = \frac{33423}{b^{14}} \times b^3$	Substituting a in terms of b into the equation
$b \approx 0.850$	Solving for b (done using numerical solve on the graphic display calculator, rounding to the nearest 3 significant figures)
$33423 = a \times b^{14}$	Referring to the original equation from the first coordinates
$33423 = a \times 0.850^{14}$	Substituting the obtained b value into the equation
$a = \frac{33423}{0.850^{14}}$	If $y = a \times b^x$, then $a = \frac{y}{b^x}$
$a \approx 325222$	Therefore $a \approx 325222$ rounded to the nearest whole number

By calculating the a and b values following this method using randomly selected coordinates from the data set, a mean average of the a and b values can be calculated to modify the function, obtaining a more accurate model.

Coordinates used		a value	b value
Coordinate 1	Coordinate 2		
$(14, 33423)$	$(3, 200000)$	325222	0.850
$(15, 27964)$	$(24, 2665)$	1410063	0.770
$(12, 47661)$	$(7, 106027)$	325753	0.852
$(11, 57708)$	$(28, 1505)$	610418	0.807
$(21, 4605)$	$(9, 83668)$	742962	0.785
Mean a and b values		682884	0.813

Table 3: Table showing the coordinates used to obtain a more accurate model, as well as the a and b values calculated from these coordinates. The bottom row shows the mean a and b values calculated which will be used in the modified model.

By substituting the obtained a and b values into the function $f(x) = a \times b^x$, it can be seen that the revised exponential function would be $f(x) = 682884 \times 0.813^x$. This will be **model 2** (the revised model).

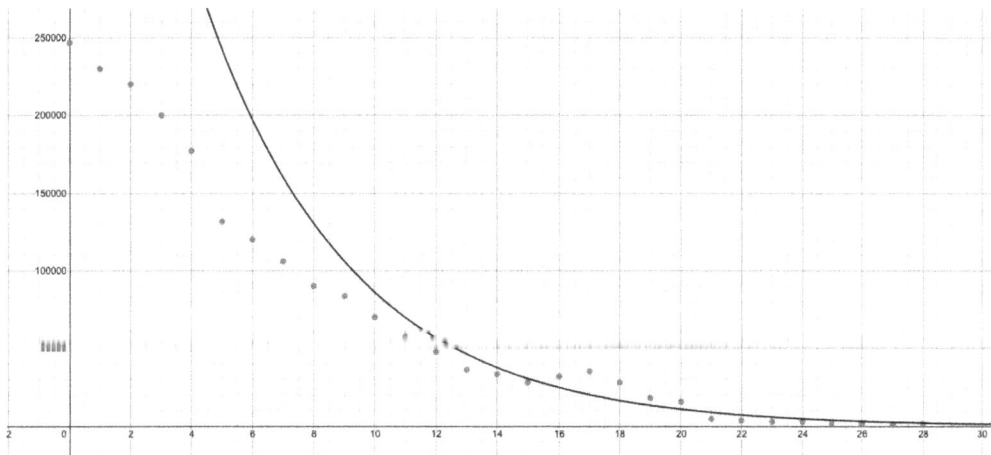

Graph 4: Model 2, $f(x) = 682884 \times 0.813^x$ plotted against the relevant data points. x-axis is years from 1990 and y-axis is cocoa production in tonnes.

Visually, this function clearly does not fit the data set as well as function 1 does. This can be seen particularly in the first 10 coordinates from the left, where the function does not come close to intersecting the coordinates from the data set. However, similar to model 1, in order to obtain the true accuracy of model 2, the expected values will be calculated and compared to the observed values, analysed using Pearson's correlation coefficient.

Eg. In 1997 (7 years after 1990), the observed amount of cocoa produced was 106027 tonnes. Therefore, the coordinate is (7, 106207). In order to find the expected value, a new y value will need to be calculated through substituting the x value into model 2.

$$y = 682884 \times 0.813^x$$
$$= 682884 \times 0.813^7$$
$$\approx 106317 \text{(rounded to the nearest whole number)}$$

x (years)	O_y (tonnes)	E_y (tonnes)
0	247000	682884
1	230000	555185
2	220000	451365
3	200000	366960
4	177172	298338
5	131475	242549
6	120071	197192
7	106027	160317
8	90183	130338
9	83668	105965
10	70262	86149
11	57708	70039
12	47661	56942
13	36236	46294
14	33423	37637
15	27964	30599
16	31937	24877
17	35180	20225
18	27955	16443
19	10152	13000
20	15654	10868
21	4605	8836

x	Oy	Ey
22	3645	7184
23	2809	5840
24	2665	4748
25	1729	3860
26	1757	3138
27	1029	2551
28	1505	2074

Table 4: Table showing the observed and expected values, rounded to the nearest whole number, obtained from model 2, to be compared using Pearson's correlation coefficient.

	r=	x	Oy	Ey	Oy-Ô	Ey-Ê	(Oy-Ô)^2	(Ey-Ê)^2	(Oy-Ô)(Ey-Ê)
2	0.96802172	0	247000	682884	177087	557271	31359866634	310551326331	98685602681
3		1	230000	555185	160087	429572	25627902771	184532115218	68768969070
4		2	220000	451365	150087	325752	22526159323	106114676013	48891268120
5		3	200000	366960	130087	241347	16922672427	58248467421	31396173867
6		4	177172	298338	107259	172726	11504530067	29834166361	18526415301
7		5	131475	242549	61562	116936	3789901072	13674126504	7198860097
8		6	120071	197192	50158	71580	2515842260	5123659079	3590308905
9		7	106027	160317	36114	34705	1304233449	1204420177	1253333588
10		8	90183	130338	20270	4725	410879889.7	22329375.63	95784609.4
11		9	83668	105965	13755	-19648	189204768.1	386036953.5	-270259194.6
12		10	70262	86149	349	-39463	121921.3746	1557348178	-13779478.6
13		11	57708	70039	-12205	-55573	148957816.4	3088379873	678261249.2
14		12	47661	56942	-22252	-68671	495143830.9	4715648197	1528045848
15		13	36236	46294	-33677	-79319	1134128716	6291464132	2671203874
16		14	33423	37637	-36490	-87976	1331507517	7739726375	3210218661
17		15	27964	30599	-41949	-95014	1759704136	9027627288	3985718640
18		16	31937	24877	-37976	-100736	1442163481	10147703818	3825525828
19		17	35180	20225	-34733	-105388	1206369312	11106585425	3660415799
20		18	27955	16443	-41958	-109170	1760459296	11918054130	4580529356
21		19	18152	13368	-51761	-112245	2679183272	12598861749	5809876044
22		20	15654	10868	-54259	-114744	2944020371	13166293924	6225900539
23		21	4605	8836	-65308	-116777	4265112344	13636827270	7626440869
24		22	3645	7184	-66268	-118429	4391424973	14025459057	7848041231
25		23	2809	5840	-67104	-119772	4502923677	14345440796	8037190119
26		24	2665	4748	-67248	-120865	4522270315	14608245752	8127880174
27		25	1729	3860	-68184	-121752	4649034344	14823664229	8301549501
28		26	1757	3138	-68156	-122474	4645216834	14999961466	8347339307
29		27	1029	2551	-68884	-123061	4744981703	15144059175	8476926548
30		28	1505	2074	-68408	-123538	4679630875	15261718273	8450988583
31	AVERAGE		69913	125613					
32	SUM/SIGMA						167453547396	917894392539	379514729737

Fig. 3: Screenshot of the calculations for model 2 done on google sheets. Cells highlighted in blue are the average of the above values and cells highlighted in green are the sum of the

above values. Cell highlighted in red is the final r-value calculated according to the equation
$$r = \frac{\Sigma(x-\bar{x})(y-\bar{y})}{\sqrt{\Sigma(x-\bar{x})^2\Sigma(y-\bar{y})^2}}.$$

According to the above calculations, it can be seen that the r-value for model 2 is equal to 0.968 when rounded to the nearest 3 significant figures. This demonstrates that there is a very strong positive correlation between the observed and expected values. As a result,t he model $f(x) = 682884 \times 0.813^x$ is an accurate exponential function to model the future fn cocoa production in Malaysia. However, this is still not as accurate as model 1 which yielded an r-value of 0.993, demonstrating a near perfect positive correlation between the observed and expected values.

The final model

Between the 2 models, it is clear that model 1 is the more accurate of the two due to its higher r-value of 0.993 compared to the r-value of 0.968 obtained from model 2. Whilst both r-values do showcase a strong positive correlation between the observed and expected values of both models, the fact that the r-value obtained from model 1 is so close to 1 makes it near impossible to disregard. Therefore, although model 1 was obtained using 2 randomly selected points from the data set, the correlation between observed and expected values is significantly higher than the function obtained from an average of a and b values, and it will thereby be used as the final model.

The exponential function $f(x) = 247000 \times 0.872^x$ will be used to model the future of Malaysia's cocoa industry overtime, specifically in the year 2050. This is because it is expected the Intergovernmental Panel on Climate Change (IPCC) predicts chocolate will be extinct by that year (Austin, 2018).

$x = 2050 - 1990$ Calculating the number of years between 1990 and 2050

$\quad = 60$ Therefore the x value of this equation is 60

$y = 247000 \times 0.872$ Function to substitute the calculated x value

$\quad = 247000 \times 0.872$ Substituting the x value into the exponential equation

$\quad \approx 66.6$ Equation solved ahd rounded to the nearest 3 significant figures

Therefore, from the exponential model $f(x) = 247000 \times 0.872^x$, it can be predicted that in 2050, cocoa production in Malaysia would reach 66.6 tonnes.

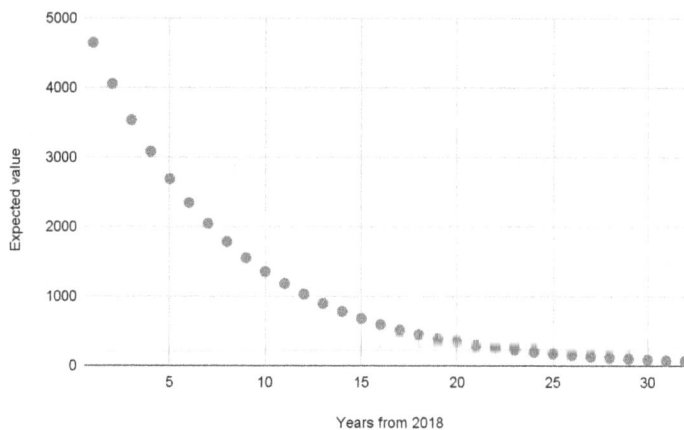

69

Graph 5: Scatter plot showing the expected values of cocoa production in toned against the years from 2018, modelled using model 1.

Conclusion

From taking the more accurate model produced in this exploration, it can be predicted that using the function $f(x) = 247000 \times 0.872^x$, cocoa bean production in Malaysia will have dropped to 66.6 tonnes per year by 2050, but will not have stopped completely. Production between 2018 and 2050 will follow a decaying exponential model, but according to this model, cocoa will not go extinct during that time period.

However, assuming that the Malaysian cocoa industry follows the trend of exponential decay, production in 2050 may not be worth it as by then, only a small amount of cocoa will be able to be harvested annually. While the theoretical model obtained in this investigation shows that cocoa production in Malaysia will never hit 0, the results suggest it may be more practical to halt cocoa production in Malaysia due to its low yield in the future. Instead, cocoa should further be cultivated in the top producing areas globally, such as Côte D'Ivoire, Ghana and Indonesia (worldatlas.com, 2015), rather than in Malaysia, a minor producer.

Evaluation

There are 2 main limitations that come with using an exponential function to model the future of the cocoa production in Malaysia. Firstly, the nature of the exponential function means that it never reaches 0, therefore it does not entertain the idea that cocoa will eventually go extinct. However, there have been sources that debunk the extinction hypothesis, such as USA Today who claim the article that sparked this idea "contains inaccurate information" (Wecker, 2018). Furthermore, Malaysia's tropical climate year round means that unless there are truly severe and detrimental environmental problems, the likelihood of cocoa production stopping entirely there due to climate change is virtually impossible. Therefore, this limitation does not severely affect the results of the exploration.

Secondly, the nature of an exponential function means that it does not entertain the idea that cocoa production will improve following the drop in 1990. The dependent variable of an exponential equation, in this case the tonnes of cocoa produced, continues to infinitely decrease without ever reaching 0. This implies that by modelling cocoa production with an exponential function, it is expected that the cocoa population will follow this downward trend, when the reality is it could improve overtime, as exemplified by the brief rise between the years 2005 and 2007. If I were to repeat this exploration in the future, I would create 2 different models: an exponential function modelling the downward trend, and a piecewise function modelling a potential improvement. These models will be used to predict 2 different outcomes which can both be considered when addressing the aims of the exploration. We do not have any way of knowing whether or not cocoa production in Malaysia will truly improve overtime seeing as climate change and agricultural laws are incredibly unpredictable, but having 2 models which pose 2 different outcomes could give us an insight into the paths the industry could take in the future.

Additionally, it is surprising that the initial model yielded a higher r-value than the revised model. Logically, a function calculated from an average of a and b values would be more accurate than one calculated from 2 random data points as it takes into account a greeted number of points from within the sample. However, the r-value made it clear that model 1 had to be used due to how close it was to 1. This implies that another pair of random data points could have yielded an r-value even closer to 1, and due to the sheer amount of data, it would be impossible to test all of them. Despite this, the r-value yielded from the two random data points was already so close to being a perfect linear correlation, therefore this limitation does not vastly impact my exploration. Nevertheless, if I had the chance to repeat this exploration, I would use more randomly selected data points to try to obtain more equations in order to have a variety of models with different expected values. These would yield a variety of r-values, the highest of which could be used as my final model.

Works cited

Austin, C. (n.d.). Is Chocolate Going Extinct? [online] Food Network. Available at: https://www.foodnetwork.com/fn-dish/news/2018/1/is-chocolate-going-extinct-#:~:text =According%20to%20scientists%20at%20the [Accessed 15 Mar. 2021].

Encyclopedia Britannica. (n.d.). Exponential function | mathematics. [online] Available at: https://www.britannica.com/science/exponential-function.

Hease, M., Humphries, M., Sangwin, C.J., Vo, N. and Al, E. (2019). Mathematics : applications and interpretation SL. 2 : for use with IB diploma programme. Marleston: Hease Mathematics.

Our World in Data. (n.d.). Cocoa bean production. [online] Available at: https://ourworldindata.org/grapher/cocoa-bean-production?tab=chart&time=earliest..20 18 [Accessed 15 Mar. 2021].

Santa Barbara Chocolate. (2017). FDA Standard Definition of Chocolate. [online] Available at: https://www.santabarbarachocolate.com/blog/fda-standard-definition-of-chocolate/#:~:t ext=To%20meet%20the%20FDA%20standard [Accessed 29 May 2021].

2. MODELING THE RELATIONSHIP BETWEEN SCORES FOR WOMEN'S ECONOMIC OPPORTUNITY INDEX

Author: Anonymous
Moderated Mark: 18/20
Level: Math AI SL

Introduction

Learning about the KOF Globalisation Index (KOFGI) in my Geography class raised the question of how a country's globalisation score would be impacted by more gender equality in the workplace. KOFGI measures a country's globalisation through three dimensions: economic globalisation, political globalisation, and social globalisation. The indicators within these dimensions include trade, tariffs, debt, tourism, migration, high technology exports, internet access, press freedom, gender parity, civil liberties, and international treaties, NGOs and organisations. (KOF Swiss Economic Institute, 2020) Most of the indicators are part of the economic dimension. Women's Economic Opportunity Index, on the other hand, was created by the Economist Intelligence Unit and measures women's access to economic opportunity in a country through numerous indicators including labour policy and practice, access to finance, education and training, women's legal and social status, and general business environment. (Economist Intelligence Unit, 2010)

This topic drew my attention as I will be studying Geography at university and statistics will be a part of my course. Globalisation is also one of my favourite phenomena in Geography, along with studying the seventeen Sustainable Development Goals (SDG). WEOI fits under SDG 5, which is Gender Equality. (United Nations Department of Economic and Social Affairs, 2015) Since the KOFGI is closely linked to economic development due to the significant economic dimension, I wanted to investigate whether there was statistical evidence for stronger economic development by providing better economic opportunities for women.

Aim

This investigation aimed to explore the relationship between scores for Women's Economic Opportunity Index (WEOI) and KOF Globalisation Index (KOFGI) for 126 countries in 2012

which the identified global pattern could be applicable to individual countries or regions. This investigation demonstrated the merits of using Pearson's correlation coefficient and linear regression in tracking progress of sustainable development, specifically the role of women, with globalisation and in predicting how certain countries that score on WEOI may score on KOFGI. Furthermore, this investigation established the convenience of technology in the mathematics that were used whilst still demonstrating how and why individual calculations led to the final results.

Methodology

The data was collected from two separate databases: KOFGI data was taken from a database produced by the Swiss Economic Institute and WEOI data was taken from Our World in Data which sourced their database from the Economist Intelligence Unit. (KOF Swiss Economic Institute, 2020; Our World in Data, 2012) These sources were chosen because of their strong, international reputation which increases the reliability of this investigation's results. Convenience sampling was used to collect the data within the databases because some countries had data for WEOI but not KOFGI, or vice versa. Convenience sampling was also used because the only most recent year where there was data for both indexes was 2012. The benefits of convenience sampling for this investigation is that the data was easier to acquire because it was readily available and not time-consuming, and the maximum amount of data could be selected so that the results would be applicable to more countries. Having the maximum number of countries included in this study also meant that a more representative correlation coefficient could be obtained.

A scatter graph with a line of best fit was initially plotted to have a visual representation of the relationship between WEOI and KOFGI scores. This would also prove to be useful when understanding the reasons for the results by identifying anomalies that were farther away from the line of best fit. Pearson's correlation coefficient was then found between WEOI and KOFGI scores to find their immediate relationship. The same was done for Asian countries only, to identify anomalies and determine whether the pattern identified globally could still be applicable to one region. After doing so, linear regression was used to interpolate data for a country with a missing KOFGI score.

Hypotheses

Null hypothesis (H$_0$): there is no correlation between WEOI and KOFGI scores for 126 countries in 2012.

Alternative hypothesis (H$_1$): there is a strong positive correlation between WEOI and KOFGI scores for 126 countries in 2012.

Understanding Pearson's correlation coefficient and linear regression

Pearson's correlation coefficient measured the strength and direction of the relationship between KOFGI and WEOI scores. (QuestionPro, n.d.) The strength of the relationship between the two variables was determined by how consistently one variable changed with the other variable. (QuestionPro, n.d.) The result — the correlation coefficient or r value — could be any value between -1 and 1, where the closer it is to -1 or 1, the stronger the relationship would be between the variables; a correlation coefficient of 0 would reflect no relationship between the variables. (QuestionPro, n.d.) The direction of the relationship, on the other hand, was identified by the slope of the line of best fit, which could either indicate a positive relationship, a negative relationship, or no relationship if it were parallel to the x-axis. (QuestionPro, n.d.)

Linear regression produced a line of best fit based on the mean of the x and y values, and was especially useful with interpolation of data since the linear regression line had an equation in the form of y=mx+c, whereby an x value (the independent variable, which was WEOI score, in this case) could be substituted into the equation to predict the y value (the dependent variable, which was KOFGI score), or vice versa. (Wong, 2020) Choosing linear regression was particularly appropriate for the context of this investigation because there was an opportunity to suggest a country's WEOI score if only their KOFGI score was available for 2012. This interpolation of data would not be possible using solely Pearson's correlation coefficient.

Exploration

Exploring Pearson's correlation coefficient in all 126 countries

A scatter graph with the WEOI and KOFGI scores was plotted, as shown in figure 1. Pearson's was appropriate since the relationship was shown to be linear without significant outliers that could drastically impact the correlation coefficient. Furthermore, both variables consisted of continuous data as opposed to discrete data, which supports the choice to use Pearson's rather than Spearman's which ranks the data instead. It was also important for observations of the data to be paired observations, so by gathering both WEOI and KOFGI scores for all 126 countries, this was made possible since, for every observation of WEOI score, there was a corresponding KOFGI score. (Magiya, 2019)

Figure 1

Investigating the relationship between WEOI ~~scores and KOFGI scores in~~

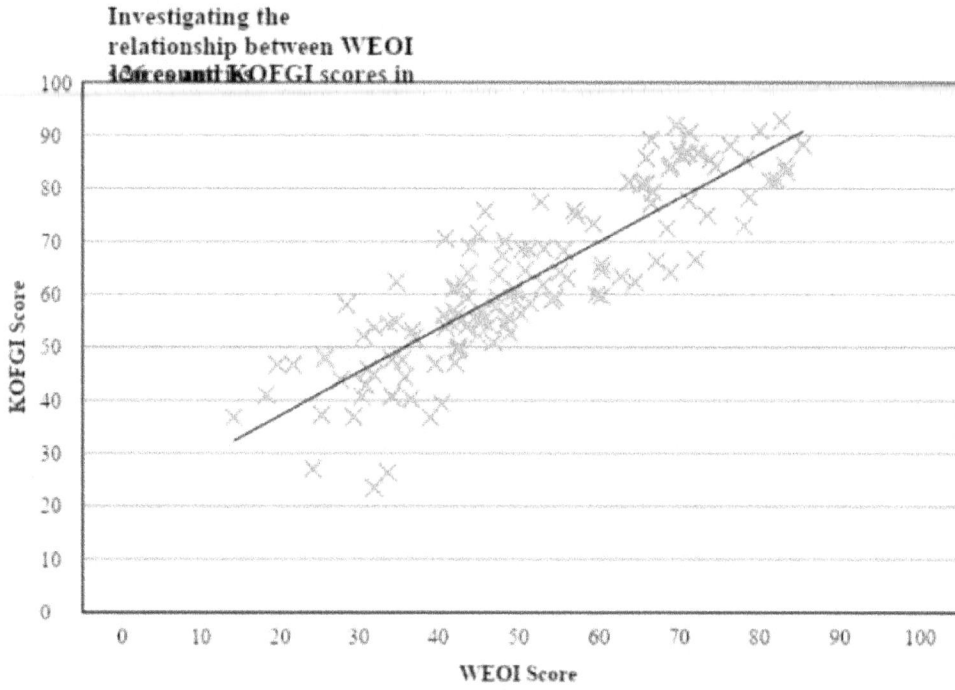

Figure 2 demonstrates Pearson's correlation coefficient equation which calculated the correlation coefficient for this investigation. This was used to represent the methodology.

Figure 2: Pearson's correlation coefficient equation (Jovancic, 2020)

$$r_{xy} = \frac{\sum_{i=1}^{n}(x_i - \bar{x})(y_i - \bar{y})}{\sqrt{\sum_{i=1}^{n}(x_i - \bar{x})^2}\sqrt{\sum_{i=1}^{n}(y_i - \bar{y})^2}}$$

Defining key terms:

r_{xy} : the correlation coefficient of WEOI and KOFGI scores

$\sum_{i=1}^{n}$: the sum of the following numbers, where i represents the index of summation that begins at the first country in the data and n is the last value of the data

x_i : the WEOI score of a country

$\bar{x}$: the mean WEOI score of all the data

y_i : the KOFGI score of a country

$\bar{y}$: the mean KOFGI score of all the data

Figure 3 represents the first 10 countries used in this investigation. This sample of 10 countries was used to show the methodology that was applied to the rest of the countries in the investigation. Although the correlation coefficient for the sample of countries in figure 3 is calculated manually, to save time, a calculator was used to calculate the correlation coefficient of the 126 countries.

Figure 3: Sample of first 10 countries in the data with calculations finding the correlation coefficient

Country	*x* WEOI Score	*y* KOFGI Score
Albania	56.473900	58.43
Algeria	39.656799	54.88
Argentina	59.150902	58.94
Armenia	53.324100	54.27
Australia	87.132698	81.60
Austria	76.291801	90.55
Azerbaijan	46.831799	56.92
Bahrain	48.981602	68.83
Bangladesh	39.167599	40.73
Belarus	53.838600	52.68

The mean value for WEOI and KOFGI were found using the following formulas:

$$mean\ WEOI = \frac{\Sigma x}{n} \quad \text{and} \quad mean\ KOFGI = \frac{\Sigma y}{n}$$

Example for Albania in the 10 sample of 10 countries:

$$mean\ WEOI = \frac{56.473900+39.656799+59.150902+53.324100+87.132698+76.291801+46.831799+48.981602+39.167599+53.838600}{10}$$

$mean\ WEOI = 56.084980$

$$mean\ KOFGI = \frac{58.43+54.88+58.94+54.27+81.60+90.55+56.92+68.83+40.73+52.68}{10}$$

$mean\ KOFGI = 61.78$

Pearson's correlation coefficient with calculation explained	Calculations with data from the sample of 10 countries				
	Albania	Algeria	Argentina	Armenia	Australia

$(x_i - \bar{x})$ Example for Albania: 56.473900 – 56.084980 = 0.388920	0.388920	-16.428181	3.065922	-2.760880	31.047718
	Austria	**Azerbaijan**	**Bahrain**	**Bangladesh**	**Belarus**
	20.206821	-9.253181	-7.103378	-16.917381	-2.246380
$(y_i - \bar{y})$ Example for Albania: 58.43 – 61.78 = -3.353	**Albania**	**Algeria**	**Argentina**	**Armenia**	**Australia**
	-3.353	-6.903	-2.843	-7.513	19.817
	Austria	**Azerbaijan**	**Bahrain**	**Bangladesh**	**Belarus**
	28.767	-4.863	7.047	-21.053	-9.103
$(x_i - \bar{x})(y_i - \bar{y})$ Example for Albania: 0.388920 + -3.35 = -2.96108	**Albania**	**Algeria**	**Argentina**	**Armenia**	**Australia**
	-1.30404876	113.403733443	-8.716416246	20.74249144	615.272627606
	Austria	**Azerbaijan**	**Bahrain**	**Bangladesh**	**Belarus**
	581.289619707	44.998219203	-50.057504766	356.161622193	20.44879714
$\sum\limits_{i=1}^{n} (x_i - \bar{x})(y_i - \bar{y})$ Adding all the values from the row above	-1.30404876+ 113.403733443+ -8.716416246 + 20.74249144 + 615.272627606 + 581.289619707 + 44.998219203 + -50.057504766 + 356.161622193 + 20.44879714 = 1692.23914096				
$\sum\limits_{i=1}^{n} (x_i - \bar{x})^2$ Adding all the squared WEOI values	$0.388920^2 + -16.428181^2 + 3.065922^2 + -2.760880^2 + 31.047718^2 + 20.206821^2 + -9.253181^2 + -7.103378^2 + -16.917381^2 + -2.246380^2$ = 2086.65847438642				
$\sqrt{\sum\limits_{i=1}^{n} (x_i - \bar{x})^2}$ Square rooting the total from the row above	$\sqrt{2086.65847438642}$ = 45.6799570313548				
Pearson's correlation coefficient with calculation explained	**Calculations with data from the sample of 10 countries**				
$\sum\limits_{i=1}^{n} (y_i - \bar{y})^2$ Adding all the squared KOFGI values	$-3.353^2 + -6.903^2 + -2.843^2 + -7.513^2 + 19.817^2 + 28.767^2 + -4.863^2 + 7.047^2 + -21.053^2 + -9.103^2 = 1943.07801$				
$\sqrt{\sum\limits_{i=1}^{n} (y_i - \bar{y})^2}$ Square rooting the total from the row above	$\sqrt{1943.07801}$ = 44.0803585511733				
$\dfrac{\sum\limits_{i=1}^{n}(x_i - \bar{x})(y_i - \bar{y})}{\sqrt{\sum\limits_{i=1}^{n}(x_i - \bar{x})^2}\sqrt{\sum\limits_{i=1}^{n}(y_i - \bar{y})^2}}$	$\dfrac{1692.23914096}{45.6799570313548 \times 44.080358551173}$ = 0.840409456939849		***Reflection*** The bigger the top value, the higher the correlation coefficient. So, the difference between the WEOI and KOFGI scores from their mean determines the strength of the correlation.		

	= 0.840 (3 s.f.)

Therefore, after establishing that the scatter graph in figure 1 demonstrated a positive, linear relationship, the calculations shown in figure 3 were applied to all 126 countries. The result was a correlation coefficient of 0.868 (3 s.f.), which is a strong positive correlation because it is quite close to 1. This means that, generally, individual data values for WEOI and KOFGI were not far from the WEOI and KOFGI mean values, suggesting that the WEOI and KOFGI values in this sample were quite similar to each other. This is supported by the cluster of data points around the middle of the graph in figure 1. Furthermore, the strong positive correlation coefficient implies that as WEOI scores increase, KOFGI scores also increase, further suggesting that gender equality can improve a country's globalisation networks.

Therefore, these findings support the alternative hypothesis of this investigation, stating that there is a strong positive correlation between WEOI and KOFGI values. The null hypothesis can therefore be rejected.

However, there were still anomalies visible in figure 1. Furthermore, Pearson's correlation coefficient does not take into account whether certain regions of the world that had more countries but similar characteristics, such as historical links to other countries, thus linking to economic globalisation, would 'carry' the correlation coefficient making it higher. Therefore, this led the exploration of this investigation to go beyond the global pattern and examine whether Pearson's correlation coefficient was more strongly influenced by some regions over others because there was another confounding variable that influenced KOFGI scores instead of WEOI, or a variable that heavily determined WEOI score.

So, the countries were divided into different regions of the world through stratified sampling: Europe; Middle East, N Africa, and Greater Arabia; Asia; North, Central, and South

America — these were originally separate, but there were too few countries in each region to consider them separately —; Australia & Oceania; and Sub-Saharan Africa.

Exploring Pearson's correlation coefficient in Asian countries

Asian countries were used to determine whether the global pattern fit the countries in Asia. This region was used because my studies in my Geography class have taught me that there is a lot of economic, social and political variation between Asian countries. There are countries that are highly globalised with a reputation of strong gender equality, whilst there are other countries that are low-income countries that are not associated with a high globalisation.

Figure 4 illustrates a scatter graph with a line of best fit indicating the strength and direction of the relationship between WEOI and KOFGI scores in Asian countries. It reflects a strong, positive relationship between the two variables, suggesting that, generally, as WEOI scores increase in Asian countries, KOFGI scores will also increase. However, there were some anomalies identified, highlighted in red in figure 4. These countries were Pakistan, Malaysia, Singapore, Japan, South Korea and Laos. After, the CIA World Factbook was used to identify characteristics related to their economies, politics, and society to suggest a qualitative reason or confounding variable that may explain why they were anomalies. (CIA, 2021) Initially, it was difficult to find a common characteristic between all the countries that suggested a reason for their deviation from the line of best fit. So, a deeper look was taken into identifying how they were anomalies. Figure 4 shows that Singapore, Malaysia, and Pakistan score higher in KOFGI than they should according to the line of best fit. Japan, South Korea, and Laos score lower in KOFGI than they should according to the line of best fit. Therefore, this suggests that for these countries, a higher WEOI score may not necessarily equate to a higher KOFGI score or that a lower WEOI score does not necessarily equate to a lower KOFGI score. Pakistan, Malaysia, and Singapore, therefore, may have a common characteristic that makes them more globalised but not allow them to score more highly on gender

82

equality in the workplace. Japan, South Korea, and Laos, on the other hand, may have a common characteristic that makes them have more gender equality in the workplace but be less globalised.

Figure 4

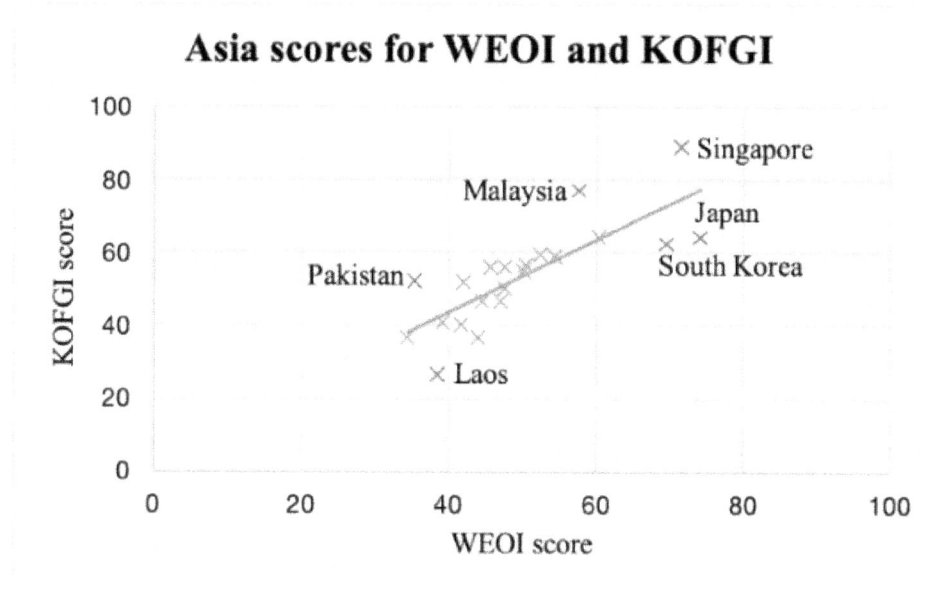

Asia scores for WEOI and KOFGI

After reviewing the data on CIA World Factbook, it was found that the countries that score higher in KOFGI were colonies of the UK. Malaysia is a former colony of the UK, Singapore used to be part of Malaysia, and Pakistan used to be part of British India. (CIA, 2021) The historical links with the UK may have contributed to the higher globalisation index because potential economic, social and political links were already established. Furthermore, the lower WEOI score can be due to both Malaysia and Pakistan having a major Muslim population, which can impact the extent to which women can participate in the local economy, thus affecting WEOI since it relies on women's opportunity to access the local economy, either through jobs or finance. (CIA, 2021) For Japan, South Korea, and Laos, South Korea used to be a part of Japan, which suggests a reason for their similar scores. (CIA, 2021) Pearson's correlation coefficient was then calculated for Asian countries, which resulted in 0.794 (3 s.f.), showing quite a strong positive correlation between WEOI and KOFGI, but lower than the global correlation coefficient.

When the global correlation coefficient was recalculated without the data from Asian countries, the correlation coefficient was found to be 0.773 (3 s.f.). This is almost 0.1 lower than the global correlation coefficient, thus suggesting that WEOI does not affect KOFGI in Asia to the same extent that it affects the 126 countries as a whole.

Exploring linear regression to predict Myanmar's WEOI score

On both graphs plotted in figure 1 and figure 4, a line of best fit was drawn. This line has an equation because it is a linear regression line. After identifying Pearson's correlation coefficient and how it produced different results based on which data was selected, the strong positive correlation coefficient and visible linear relationship on the graphs suggested that it was appropriate to interpolate data.

Only 126 countries were used in this investigation because sometimes there was data missing for WEOI. For example, Myanmar was not included in this investigation because there wasn't available data for its WEOI score. However, through linear regression that has a line with the equation in the form of $y = mx + c$, the y value (or KOFGI value) of Myanmar could be substituted into the equation to suggest what the WEOI may be according to the trend identified in the 126 countries. Using a calculator, the equation of the line was found to be $y = 0.8219x + 16.53$. However, in order to interpolate the domain, the y value needed to be within the realm of the domain. So, the KOFGI value could not be lower than 23.44 or higher than 92.76. This is because there was no evidence in this investigation to suggest that the trend identified in figure 1 continued after 92.76 or occurred before 23.44. In other words, extrapolation could not occur. So, the following calculation was done to interpolate the WEOI score for Myanmar:

$$33.57 = 0.8219x + 16.53$$

$$17.04 = 0.8219x$$

$20.7 \ (3 \ s.f.) = x$

Myanmar is also an Asian country, so by categorising the data by region, as was done to find differences in Pearson's correlation coefficient, the same could be done to interpolate data most reliably. Therefore, the equation of the linear regression line in figure 4 was found to be $y = 0.9945x + 4.1485$. Myanmar's KOFGI score of 33.57 was substituted into figure 4's regression line equation to reveal an interpolation of 29.6 (3 s. f.) as the potential WEOI score for Myanmar. This is more reliable than using the potential WEOI value calculated based off of figure 1's regression line equation because Myanmar may share common characteristics with other Asian countries as was seen when finding common characteristics between Pakistan, Singapore, and Malaysia.

Conclusion

In conclusion, Pearson's correlation coefficient has been particularly useful in identifying the global trend in WEOI and KOFGI scores, however, to increase the reliability of the results, the data needed to be categorised to examine the relationship in that category. Asia was the category used to delve into the relationship between WEOI and KOFGI scores, and it showed that there can be other confounding variables, such as historical colonial links, that can influence a country's KOFGI score. In addition, it reflected one of the limitations of using Pearson's correlation coefficient as it does not suggest a reason for there being anomalies. Furthermore, it showed how a group of data that shares a common characteristic, such as South Korea being a former colony of Japan, can score similarly and thus more heavily influence the linear regression line in the direction that they are plotted on the graph. However, perhaps the biggest limitation of Pearson's correlation coefficient is that the results of this investigation are merely correlational; no causal relationship between WEOI and KOFGI scores can be inferred. By including qualitative data, as was done for

Asian countries, other reasons helped justify the correlational relationship, but this was still not enough to infer causation.

Linear regression, however, was especially beneficial in providing an indication of the direction of the relationship, demonstrating that it was positive. It further allowed missing data to be interpolated which is extremely useful in predicting data to track sustainable development in relation to globalisation and gender equality. However, it shares some of the limitations that Pearson's correlation coefficient has, such as not being able to infer a cause-effect relationship between WEOI and KOFGI and offering no justification to the gradient of the line. Yet, this emphasises the benefits in the methodology of this investigation because these precautions were taken, supporting the reason for qualitative characteristics to be researched in the sample of Asian countries.

Another strength of this investigation is the large sample size. 126 countries were chosen through convenience sampling ensuring that the largest number of countries were chosen, which grants the results of this investigation validity to a lot of countries. Justifications for the relationship between WEOI and KOFGI in Asian countries could not have been executed had stratified sampling not been chosen to categorise the data in the exploration. However, the countries included in this investigation had data for both WEOI and KOFGI, which introduced a potential bias in data collection since it is possible that countries that have a high score for one variable may naturally have a high score for the other variable. For instance, WEOI is reliant on there being lots of economic opportunities and resources for women, which can be more prominent in a country that is more globalised and having more diverse economic opportunities.

Nevertheless, this investigation was able to suggest that improving women's economic opportunities can lead to higher globalisation scores, which could benefit a country's economic growth and role in geopolitics. Furthermore, this investigation demonstrated the merits of applying Pearson's correlation coefficient, linear regression, and sampling techniques to better understanding how gender equality may impact globalisation and how this mathematics helps to predict a

country's score in one of the variables.

Bibliography

Asselin, S., 2015. *Making Hay: The Future of U.S. Competitiveness in the Age of Globalization.* [online] GSIPM. Available at: <https://sites.miis.edu/gsipm/2015/04/14/making-hay-the-future-of-u-s-competitiveness-in-the-age-of-globalization/> [Accessed 1 March 2021].

CIA, 2021. *Japan.* [online] Cia.gov. Available at: <https://www.cia.gov/the-world-factbook/countries/japan/> [Accessed 1 March 2021].

CIA, 2021. *Malaysia.* [online] Cia.gov. Available at: <https://www.cia.gov/the-world-factbook/countries/malaysia/> [Accessed 1 March 2021].

CIA, 2021. *Pakistan.* [online] Cia.gov. Available at: <https://www.cia.gov/the-world-factbook/countries/pakistan/> [Accessed 1 March 2021].

CIA, 2021. *Singapore.* [online] Cia.gov. Available at: <https://www.cia.gov/the-world-factbook/countries/singapore/> [Accessed 1 March 2021].

CIA, 2021. *South Korea.* [online] Cia.gov. Available at: <https://www.cia.gov/the-world-factbook/countries/korea-south/> [Accessed 1 March 2021].

CIA, 2021. *The World Factbook.* [online] Cia.gov. Available at: <https://www.cia.gov/the-world-factbook/countries/> [Accessed 1 March 2021].

Economist Intelligence Unit, 2010. *Women's economic opportunity | A new pilot index and global ranking from the Economist Intelligence Unit | Findings and methodology.* [ebook] Economist Intelligence Unit. Available at: <http://graphics.eiu.com/upload/WEO_report_June_2010.pdf> [Accessed 28 February 2021].

Economist Intelligence Unit, 2010. *Women's economic opportunity.* [online] The Economist. Available at: <http://graphics.eiu.com/upload/WEO_report_June_2010.pdf> [Accessed 18 February 2021].

Jovancic, N., 2020. *Pearson Correlation Coefficient Formula: How to Calculate and Interpret.* [online] LeadQuizzes. Available at: <https://www.leadquizzes.com/blog/pearson-correlation-coefficient-formula/> [Accessed 28 February 2021].

KOF Swiss Economic Institute, 2020. *2020 KOF Globalisation Index: Variables description.* [ebook] KOF Swiss Economic Institute. Available at: <https://ethz.ch/content/dam/ethz/special-interest/dual/kof-dam/documents/Medienmitteilungen/Globalisierungsindex/KOFGI_2020_variables.pdf> [Accessed 28 February 2021].

3. INVESTIGATING THE RELATIONSHIP BETWEEN SCORING AVERAGE AND PUTTS PER ROUND AVERAGE ON THE PGA TOUR

Author: Aayush Vyas
Moderated Mark: 19/20
Level: Math AI SL

Introduction

A golf player myself, I want to investigate importance of putting for golf scores: an exploration into this topic would help me determine what to put emphasis on during my practise sessions. Thus, this exploration will investigate the relationship between scoring average (average total number of strokes per round) and putts per round average (average number of short shots made on the green around the hole) among players of the PGA Tour. The goal of this exploration is to determine the strength and nature of the relationship between scoring average and putts per round average and establish whether a player's putting abilities are independent of his golf scores.

It is expected there will be a weak positive relationship between the variables: the reason for that is that better players, so players with better scoring averages, perform better in a variety of statistics, and a putts per round average is only one of them.

In order to collect data, the list of players who were classified by the PGA Tour was gathered and organised according to their putts per round averages in the season 2019. Their scoring averages in the same season were also collected. A website with the official statistics of the PGA Tour was used as a source of the data.

Procedure of investigation

1.   Collecting data: sampling and organising data into a table.

2.   Performing the chi-squared test for independence.

3.   Plotting a scatter plot, finding the line of the best fit and analysing the Pearson's correlation coefficient of the linear model.

4.   Finding the line of quadratic regression and analysing the Pearson's correlation coefficient of the quadratic model.

5.   Drawing conclusions.

6.   Discussing areas for possible improvement and further research.

Data collection

   To collect the data, ranks of players of the PGA Tour by putts per round average and by scoring average in the season 2019 were used. For sampling, the list of players by putts per round average was chosen instead of scoring average because there were 37 players more on the scoring average rank. Avoiding a situation in which a player would have to be replaced because his putts per round average would not be available was desirable, as that would have disrupted the sampling system and would have led to less reliable results. The sample for the investigation was chosen using systematic sampling (meaning that players were selected starting at a random position with a fixed periodic interval). Putts per round averages were available for 190 players and it was decided that a sample of 95 would be chosen, which meant that the interval was two. In numerous instances, there were a few players with the same putts per round average; when that was the case, a player from those with the same putts per round averages was chosen at random. Scoring averages were not taken into consideration when making this decision, because that could have made the sample biased. If, for example, the player with the highest scoring average was always chosen, the sample would underestimate the strength of the relationship if it were positive, and overestimate its strength if the relationship were negative.

   When filling in data about scoring averages, two instances in which scoring average was unavailable for a player were encountered. In these cases, deterministic hot deck imputation was applied. Deterministic hot deck imputation is a method used for handling missing data in which a missing value is replaced with an observed value from the most similar unit (Andridge and Little 40). The unit for which data is inserted is called a recipient while that from which the data is being taken is called a donor. To perform deterministic hot deck imputation, the player whose putts per round average was closest to that of the recipient was taken as donor (in both instances in this investigation that meant choosing another player with the same putts per round average).

Sample entries (top two and bottom two players, and one player for which deterministic hot imputation was performed) are available in table 1 for illustrative purposes. Full dataset used is available in Appendix A.

Table 1

Selected players of the PGA Tour, their scoring averages and putts per round averages in the 2019 season

Note: Blue colour indicates players for which deterministic hot deck imputation was performed. Names of donors are in parenthesis.

Name	Scoring average	Putts per round average
Jordan Spieth	71.46	27.71
Justin Rose	71.82	27.94
Trey Mullinax (J. J. Spaun)	72.24	29.29
John Chin	72.49	30.13
Corey Conners	70.78	30.17

Sources of data: "Putts per round." *PGA Tour Statistics.* PGA Tour,

www.pgatour.com/content/pgatour/stats/stat.119.y2019.eoff.t060.html. Accessed 15

Apr. 2020 and

"Scoring average." *PGA Tour Statistics.* PGA Tour, www.pgatour.com/stats/stat.120.html.

Accessed 15 Apr. 2020

Chi-squared test

Firstly, the chi-squared test was performed to establish whether the correlation between the variables is statistically significant, i.e. unlikely to be a result of chance. For the purposes of this test:

H_0: scoring average is independent of putts per round average.

H_1: scoring average is dependent on putts per round average.

Players were grouped based on means of the two data sets - one of scoring averages and one of putts per round averages (Table 2). This constitutes the observed frequencies – actual number of times each event occurred.

Table 2

Observed frequencies

	Below mean putts per round	Above mean putts per round	Total
Below mean scoring	29	22	51
Above mean scoring	15	29	44
Total	44	51	95

Expected frequencies (how many times each event is expected to occur if there were no relationship between the variables) were calculated by multiplying the observed value in the given scoring category and the given putting category, and then dividing the product by the total number of players. For example, the expected value (f_e) of the number of players with scoring average below mean and putts per round average below mean:

$$f_e = \frac{44 \times 51}{95} = 23.6210 \ldots \approx 23.62 \text{ (2 DP)}$$

Table 3

Expected frequencies

	Below mean putts per round	Above mean putts per round	Total
Below mean scoring	23.62	27.38	51
Above mean scoring	20.38	23.62	44
Total	44	51	95

To interpret the results of a chi-squared test, one needs to know the number of degrees of freedom and decide what significance level they will use. The significance level represents the probability of rejecting a null hypothesis that is true (for example because the result of the chi-squared test obtained is a result of chance alone). A significance level of 0.05 was assumed. This is a fairly low probability of rejecting a true null hypothesis and therefore a significance level of 0.05 was deemed appropriate.

To interpret the results of the chi-squared test, the number of degrees of freedom was calculated using the following formula:

Degrees of freedom = (number of columns-1) × (number of rows-1)

In the present case:

Number of columns - 2

Number of rows - 2

Therefore, the number of degrees of freedom equals 1. Thus, Yates's correction for continuity will be applied in the chi-squared test, as it allows for a more precise interpretation of the results of the test in small datasets, such as if the number of degrees of freedom is 1 (Yates 217). The formula for the chi-squared with Yates's correction for continuity test is as follows:

$$\chi^2_{Yates} = \sum_{i=1}^{n} \frac{\left(\left|O_i - E_i\right| - 0.5\right)^2}{E_i}$$

Where:

O_i - observed frequency

E_i - expected frequency

n - number of distinct events

After inserting the data into the above formula and doing the initial calculations, the following equation was received:

$$\chi^2 = 1.01 + 0.87 + 1.17 + 1.01 = 4.06$$

Comparing the results to data from a table of critical values (Appendix B), it can be seen that χ^2_{calc} is greater than χ^2_{crit}. This means that the null hypothesis should be rejected, which leads to the acceptance of the alternative hypothesis - that scoring average is dependent on putts per round average. However, if the chi-squared test were to be conducted at a 0.025 significance level, χ^2 would be smaller than χ^2_{crit}, which means the null hypothesis that scoring average is independent from putts per round average would have to be accepted. Since the alternative hypothesis can only be accepted at a 0.05 significance level, but not at a 0.025 one, it can be concluded that there is a correlation between scoring average and putts per round average, but it is not very strong.

After the relationship had been confirmed to be statistically significant, its nature was investigated.

Analysing the relationship

Line of best fit

A scatter plot with a line of best fit was created using Google Sheets to see whether there was a visible relationship between the two variables (figure 1).

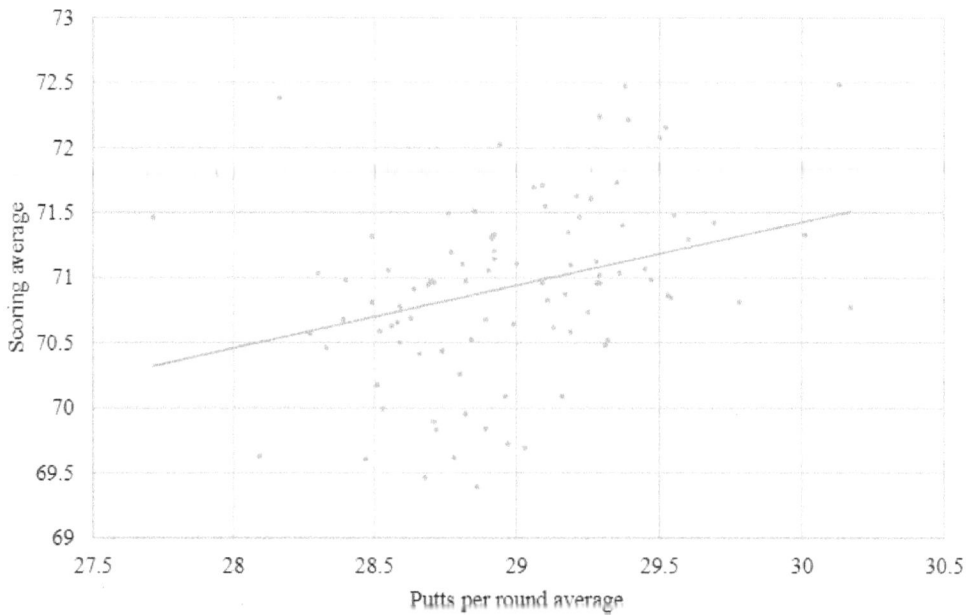

Figure 1. Scatter plot with linear regression

The equation for the line of best fit is $y = 0.48x + 56.91$, which confirms the hypothesis that the relationship is positive. The Pearson's correlation coefficient was calculated to establish the strength of the relationship. The formula is as follows:

$$r = \frac{n \times (\sum xy) - (\sum x) \times (\sum y)}{\sqrt{[n \times (\sum x^2) - (\sum x)^2] \times [n \times (\sum y^2) - (\sum y)^2]}}$$

Where:

r - Pearson's correlation coefficient

n - sample size, in the present case 95

$\sum xy$ - sum of products of the product of pairs of x and y values, in the present case 195169.52

(2 DP[1])

$\sum x$ - sum of x values, in the present case 2751.61 (2 DP)

$\sum y$ - sum of y values, in the present case 6737.96 (2 DP)

$\sum x^2$ - sum of squares of x values, in the present case 79717.47 (2 DP)

$\sum y^2$ - sum of squares of y values, in the present case 477940.39 (2 DP)

After inserting the data into the formula, the following equation was received:

$$r = \frac{95 \times 195169.52 - 2751.61 \times 6737.96}{\sqrt{(95 \times 79717.47 - 2751.61^2) \times (95 \times 477940.39 - 6737.96^2)}}$$

$$= 0.3148564452 \dots \approx 0.315$$

Since the absolute value of r is greater than 0.25 but smaller than or equal to 0.5, and r is positive, there is a weak positive correlation between the two variables (Chang Wathall et al. 272).

The value of r^2, which shows how close data points are to the line of best fit, is, to three significant figures, 0.0992. This is a very low value, as it indicates that only 9.92% of the variance of the y-variable can be explained with the x-variable. Because of how low value of r^2 is, quadratic regression was calculated. Although was is difficult to say whether quadratic regression would be more accurate in representing the dataset when looking at the graph with a bare eye, there was a possibility of it being a more appropriate fit, and thus it was calculated.

Quadratic regression

The scatter plot with quadratic regression is presented in figure 2.

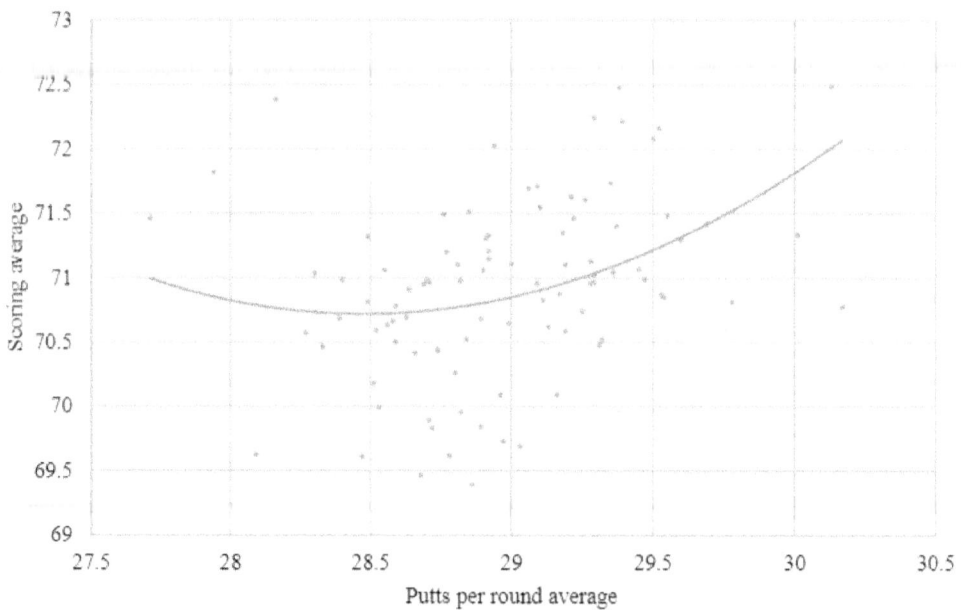

Figure 2. Scatter plot with quadratic regression

The equation of the line of quadratic regression is $y = 0.47x^2 - 26.72x + 451.15$, which also indicates a positive relationship. The r^2 was calculated using a TI-84 Plus graphing calculator to establish whether the quadratic model represents the dataset more accurately than the linear model. The value of r^2 is, to three significant figures, 0.143, which means the quadratic regression represents the trend in the data set better than the linear regression.

The fact that the vertex of the curve located at (28.47, 70.71) is interesting, as that suggests that players with putting averages lower than that have increasing scoring averages. Although this is true for the data set, it seems counterintuitive empirically. This effect can be explained by the low availability of players with putting averages lower than the x-value of the vertex. Only 9 players in the sample have putting averages lower than 28.47 (the x-coordinate of the vertex), which is a small number, and that could have led to the distortion of the quadratic regression model. Thus, the linear model can be deemed more realistic.

Conclusion

As expected, the relationship between putts per round averages and scoring averages turned out to be weak and positive; this was already visible on the scatter plot and calculations (the chi-squared test and both regressions) proved it. Three statistical operations performed (the chi-squared test and calculating the r^2 value for a linear and quadratic model) all showed that there is a weak correlation between the two variables. The Pearson's Correlation Coefficient was equal to 0.315, which means there is a weak positive correlation. The chi-squared test conducted at a 0.05 significance level resulted in accepting the alternative hypothesis that the variables are dependent on each other. The fact that if the chi-squared test were conducted at a 0.025 significance level, the result would be the opposite strengthens the conclusion that the relationship between the two variables is not strong.

As stated in the introduction, there is a wide range of aspects that determine a player's result in the form of his scoring average. Putting, measured by putts per round average, is only one of them. This can be a cause of why the relationship is only weak. It follows from the results of this exploration that practising putting, albeit important, is alone not enough to make one an outstanding player.

Mathematical processes were conducted several times, using Google Sheets, a graphing calculator (TI-84 Plus) and by hand, which served to ensure that the results are correct. Furthermore, the two methods of analysis (the chi-squared test and regression analyses) that were conducted both showed the same results, confirming that the variables under investigation are related, albeit weakly. All this means that the results of the investigation can be accepted with a high degree of confidence.

Evaluation

In order to improve the design of the study, the players for whom scoring averages were unavailable could have been crossed out before sampling was carried out from the ranking of putts per round averages (the list which was used for sampling). While the putts per rounds averages of the donor players were the same as those of the recipients, crossing out players for whom data were unavailable would have made the investigation easier to carry out, as there would have been no need to perform deterministic hot deck imputation.

To further improve the design of the investigation, putts per green in regulation instead of putts per round could have been used. On greens on regulation, which are approached from a larger distance, putts are, usually, longer than when greens are entered with a chip – from a short distance. As putts per round average is also influenced by the length of putts, using putts per green in regulation have better reflected to what extent a player's putting abilities influences his score. A player that misses many greens might have a low putts per round average not because his putting is exceptional, but because he plays many putts after chips, which tend to be shorter, and thus easier to make, than an average putt. Such a player would have a low putts per round average, even if his putting were not that good compared to others'. Putts per round average, therefore, did not isolate putting as a variable as well as putts per green in regulation would have had.

A further limitation of the study is that it only investigated male players. There are significant differences in the style of play of men and women, including that women average more putts (Rudy 2010). Further developments of this investigation might look at the trends investigated in this exploration among female players.

4. THE COASTLINE PARADOX AND FRACTAL DIMENSIONS

Author: Anonymous
Moderated Mark: 18/20
Level: Math AI SL

1. Introduction

As the threat of climate change is increasingly being emphasized by scientists and activists around the world, I began pondering about one of its many consequences, the melting of ice caps which cause rising sea levels. We all know that its effects could lead over a period of time to submerged islands. Thus, it led me to ask the question; How are coastlines affected? Climate change rushes the process by which weather and erosion impact its shapes. Yet, are their lengths impacted? Whilst, measuring the area of an island or a country is easily measured in terms of km^2. However, attempting to measure the coastline length as easily is impossible. It appears to be far more challenging that one may expect. All of these questions drove my study of how the coastlines of two islands that are prone to be submerged in a near future could be defined. After further research, I realized, as highlighted by Benoit Mandelbrot, that coastlines are better described with the use of fractal dimensions.

In this investigation, I will explore the theory of fractal dimensions in an attempt to respond to the following research question: **What is the fractal dimension of the Kiribati and Bikar islands?** Both of these islands are designated as in danger of sinking due to the impending consequences of climate change. The rising sea level effect on their coastlines highlights how ineffective is the measurement of coastline length, as its edges are evolving continuously.

2. The Coastline Paradox

In attempting to investigate how can we estimate the length of a coastline, I discerned the various problems that could arise. Can we measure it by using online softwares? Or does someone physically measure the distance on the island? Does weather and erosion affect the length of a coastline? Where would one start and what direction should one take? One would not think that Norway has a larger coastline length that Russia. However, according to CIA measurements, it is ranked at second largest due to its coastline shape ([1]). The length of coastlines is determined by approximating intricate curves as straight lines. Nevertheless, this leads to a significant issue as the length of the coastline increases as the size of the ruler decreases. The smaller the ruler becomes, the more the size of the coastlines reaches infinity as more details and complexity is figured. This is known as the Coastline Paradox.

Lewis Fry Richardson first reported this phenomenon in 1951 and noted the proportion between the length of national boundaries and scale size ([1]). In 1967, Benoit Mandelbrot elaborated on this paradox in his paper *How Long Is the Coast of Britain?* ([2]). He highlighted the difficulty in estimating coastlines:

"Seacoast shapes are examples of highly involved curves such that each of their portions can in a statistical sense be considered a reduced-scale image of the whole." ([2])

A characteristic he refers to as statistical self-similarity as regardless of the scale used to view a coastline, the overall image seems approximately the same.

"To speak of a length for such figures is usually meaningless... As even finer features are taken into account, the total measured length increases, and there is usually no clear-cut

[1] *RealLifeLore. (2018, March 3). The Coastline Paradox Explained [Video]. YouTube. https://www.youtube.com/watch?v=kFjq8PX6F7I*

[2] *Mandelbrot, B. (1967). How Long Is the Coast of Britain? Statistical Self-Similarity and Fractional Dimension. Science, 156(3775), 636-638. Retrieved November 6, 2020, from http://www.jstor.org/stable/1721427*

gap or crossover, between the realm of geography and details with which geography need

not be concerned." ([2])

For instance, if one observes the coastline of Bora Bora in Fig 1, it depicts this statistical self-similarity mentioned.

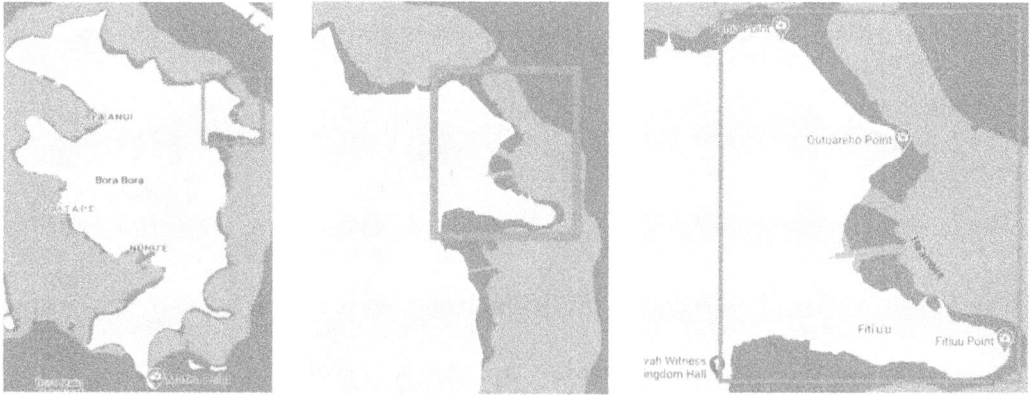

Fig 1. The island of Bora Bora at distinct magnifications to illustrate the intricacy of the coastline ([3])

From the observation of these images, as the magnification increases and further complexity is shown, the coastline does not 'smooth out'. This raises the difficulty in mapping the coastline as at whatever point is chosen, the complexity cannot be simplified as an accurate straight line. On the other hand, in Euclidian geometry, if you zoom onto a curved line it ultimately becomes a straight line that can be measured with no challenge. Therefore, theories from Euclidean geometry cannot be adequately applied to coastlines as underlined by Mandelbrot. This led to his proposal of using fractal dimension to define coastlines.

---

[3] *Google Maps. (n.d.). Google Maps. Retrieved 25 October 2020, from* https://goo.gl/maps/evcApeK63zcK9ECU9

3. Fractal Dimensions

3.1. Definition of fractal dimension

In Euclidian geometry, shapes have integer dimensions. A point is attributed to a dimension of 0, and a line has dimension 1. While an area holds a dimension of 2 and a volume has a dimension of 3. From these elements, we obtain the basic figures of traditional geometry; triangles, squares, circles, cones, cubes and spheres.

Benoit Mandelbrot defines a dimension as; n dimensional shape is one that is scaled by a factor $1/x$, the shape creates x^n copies of itself. For instance, when a cube (n=3) is scaled by a factor $\frac{1}{3}$, it creates 3^3 which results in 27 copies of itself. The power of 3 is defined as its dimension.

This concept is expressed by the following equation:

$$n = \left(\frac{1}{s}\right)^D \quad [4]$$

$s = scale\ factor$

$D = dimension$

$n = number\ of\ copies\ created$

The equation can be rearranged using log law to find D:

$$\log n = \log \left(\frac{1}{s}\right)^D$$

[4] Chapter 4: Calculating Fractal Dimensions. (2012). Wahl. Chapter 4: Calculating Fractal Dimensions. (2012). Wahl.

$$\log n = D \, \log \frac{1}{s}$$

$$D = \frac{\log n}{\log \frac{1}{s}}$$

It defies the classical perception of dimensions as integers.

3.2. Fractal dimensions of the Sierpinski triangle and Koch snowflake

To demonstrate my comprehension of fractal dimensions, I will explain how we can find the fractal dimensions of man-made fractals; The Koch Snowflake and the Sierpinski Triangle. As they are widely used to illustrate this concept. To solve for their fractal dimensions, the shapes will be sized down by a scale factor that is given, and then the shapes are counted to find the number of copies of itself that were produced.

3.3. Koch Snowflake

To construct a Koch snowflake, we start with a straight line. Then the central portion of the line is removed, and replaced with two lines with the same length of ⅓ as the outstanding lines on each side. This new form defines a rule used to create a new form repeatedly. This rule states to take each line and substitute it with four lines, each one third the length of the original. Therefore, the length of the curve increases with each repetition as it reaches infinite length.

$$Scale\ (s) = 1 \qquad\qquad Scaled\ by\ s = \frac{1}{3} \quad Number\ of\ copies\ (n) = 4$$

Fig 2. Koch snowflake ([4])

$$D = \frac{\log n}{\log 1/s}$$

$$D = \frac{\log 4}{\log 1/(1/3)}$$

$$D = \frac{\log 4}{\log 3}$$

$$D \approx 1.26$$

3.4. Sierpinski Triangle

We start with an equilateral triangle, we take the midpoint of each of the edges resulting in three triangles. This process is repeated increasing the number of triangles infinitely.

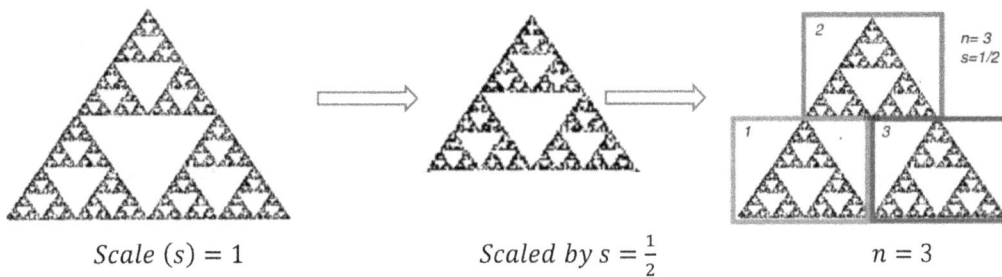

$$Scale\ (s) = 1 \qquad\qquad Scaled\ by\ s = \frac{1}{2} \qquad\qquad n = 3$$

Fig 3. Sierpinski Triangle ([5])

$$D = \frac{\log n}{\log \frac{1}{s}}$$

[5] *Fractals & the Fractal Dimension. (n.d.). Vanderbilt. Retrieved 20 October 2020, from https://www.vanderbilt.edu/AnS/psychology/cogsci/chaos/workshop/Fractals.*

$$D = \frac{\log 3}{\log \frac{1}{\frac{1}{2}}}$$

$$D = \frac{\log 3}{\log 2}$$

$$D \approx 1.58$$

4. Relationship between fractal dimensions and coastlines

"Clouds are not spheres, mountains are not cones, coastlines are not circles, and bark is not smooth, nor does lightning travel in a straight line."(Mandelbrot, 1983) ([5]). Fractal dimensions measure the complexity of geometric shapes and lines such as coastlines. As seen with the Koch Snowflake, it is a shape with a finite area but with an infinite perimeter. Coastlines have a finite area with an endless perimeter. For a coastline, the fractal dimension would be between 1 and 2 where the closer the dimension gets to 2, the more irregular is the line ([6]). For instance, the smoothest coastline is identified as South Africa with a fractal dimension of 1.02 ([7]). Therefore, rather than calculating the length of the Kiribati and Bikar islands' coastlines, I will attempt to determine their fractal dimensions. To carry out this research I will use the Hausdorff technique ([8]). Firstly, I will measure the coastline of Great Britain in order to explain how the method works. In Mandelbrot's report, he found its value at D=1.25 ([2]). This will allow me to compare my results with the set value to give a sense of the probable degree of error in applying this method. Nevertheless, there are other methods that could be used to

[6] eHowEducation. (2013, February 7). How Do Fractals Work? : Advanced Math [Video]. YouTube. https://www.youtube.com/watch?v=YiGBNNDDgH0

[7] FractalMan, W. B. (n.d.). Fractal Foundation – Fractals are SMART: Science, Math and Art! Fractal Foundation. Retrieved 16 October 2020, from http://fractalfoundation.org/

[8] Schleicher, D. (2007). Hausdorff Dimension, Its Properties, and Its Surprises. The American Mathematical Monthly, 114(6), 509-528. Retrieved November 5, 2020, from http://www.jstor.org/stable/27642249

determine the fractal dimension of coastlines. The box-counting method is a prominent technique ([7]). It involves covering the image of the island with a grid. Then count how many boxes are covering its surface. The size of the boxes is decreased repeatedly which showcases the structure of the shape more accurately. The fractal dimension is set as the slope of the line plotted from the equation $N = cs^D$ ([9]).

$$N = number\ of\ boxes$$

$$c = proportionally\ constant$$

$$s = scale\ factor$$

$$D = fractal\ dimension$$

This technique is more challenging to carry out as it requires skills in programming in order to avoid a miscount of boxes.

5. Hausdorff method

The mathematician Felix Hausdorff introduced the idea that dimensions of shapes could be non-integer values ([6]). In Euclidian dimensions objects increase by $N = r^D$ ([5]) as seen in Fig. 4. To solve for D, we take the log of both sides to get $D = log\ N/log\ r$. Thus, Hausdorff discerned that the dimension could be a fractal and developed a method to measure the edges of complex shapes. Mandelbrot used this method in his report to define the coastline of Great Britain. It consists in employing a ruler with length G and approximating multiple straight lines along the coastline. The final length of the

[9] Karperien, A. (n.d) "'Box Counting.'" National Institutes of Health, U.S. Department of Health and Human Services, imagej.nih.gov/ij/plugins/fraclac/FLHelp/BoxCounting.htm

coastline is G multiplied by the number of rulers (n) used to cover the surface. Mandelbrot established how the length of the coastline ([2]) depends on G. Being G decreases, L increases as the coastline becomes more intricate since one is able to discern finer details.

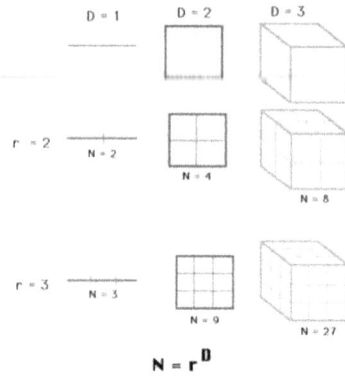

$$N = r^D$$

Fig 4. Euclidian dimensions ( )

This relationship is described by the equation:

$$L = MG^{1-D} \quad (^2)$$

$L = $ length of coastline

$M = $ proportionally constant

$G = $ length of ruler

$D - $ fractal dimension

This equation can be changed into y=mx+b form with the use of log laws:

$$L = MG^{1-D}$$

$$\log L = \log MG^{1-D}$$

$$\log L = \log M + \log G^{1-D}$$

$$\log L = \log M + (1 - D) \log G \quad (^{10})$$

Plotted as:

$$y = \log L, \quad m = (1 - D), \quad x = \log G, \quad b = \log M \quad (^{10})$$

---

[10] Lesmoir-Gordon, N., & Edney, R. (2005). Introducing Fractals: A Graphic Guide (Revised ed.) [E-book]. Icon Books Ltd. https://zbukarf1.ga/book.php?id=7PGmAwAAQBAJ

Therefore, the equation for the fractal dimension is:

$$D = 1 - m \ (^2)$$

5.1. <u>Hausdorff method to calculate Great Britain's coastline</u>

Fig 5. ([11])			
Length of ruler (G) (km)	200	50	25
log (G)	2.301	1.699	1.398
Number of rulers (n)	18	72	169
Length of coastline (km) $(L = G * n)$	2700	3600	4225
log (L)	3.431	3.556	3.626

To determine the slope (m), the data is then plotted on a graph as log L against log G:

---

[11] Gurung, Kris. (2017). Fractal Dimension in Architecture: An Exploration of Spatial Dimension.

Great Britain's coastline

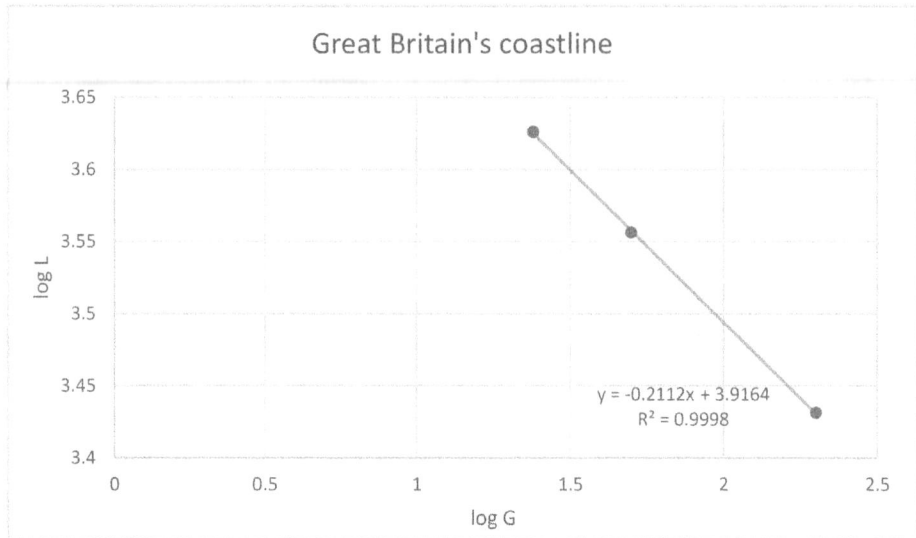

Graph of log L against log G for Great Britain's coastline, with trendline $y = -0.2112x + 3.9164$ and $R^2 = 0.9998$.

Calculating fractal dimension of Great Britain

$$D = 1 - m$$

From the graph,

$$m = -0.2112$$

$$D = 1 - (-0.2112)$$

$$D = 1.2112$$

$$D \approx 1.21$$

5.2. Evaluation of the Hausdorff method

The value I obtained was close to Mandelbrot's value of 1.25 ([2]). It holds a difference of 0.04. This appears as a 3.8% error. This error is caused by the placement of rulers. As they are placed on the coastline by eye, it is only an approximation. Since the table used was from an external source, the placement of rulers can differ in location depending on the individual that carries out this investigation. Therefore, the number of rulers that can fit on the coastline can vary, causing a different fractal dimension. Ultimately, the error of 3.8% cannot be easily avoided within the method pursued. To support this claim, the r^2 factor of 0.9998 implies that the position of rulers from Fig 5 was precise and that the method was followed correctly. However, the percentage error obtained from the calculation of Britain's fractal dimension will be used in my evaluation of the results obtained for the fractal dimensions of the Kiribati and Bikar islands.

6. Calculating the fractal dimensions of the Kiribati island

Table 1			
Length of ruler (G) (km)	150	50	25
log (G)	2.176	1.699	1.298
Number of ruler (n)	23	73	178
Length of coastline (km) $(L = G * n)$	3450	3650	4450

115

log (L)	3.538	3.562	3.648

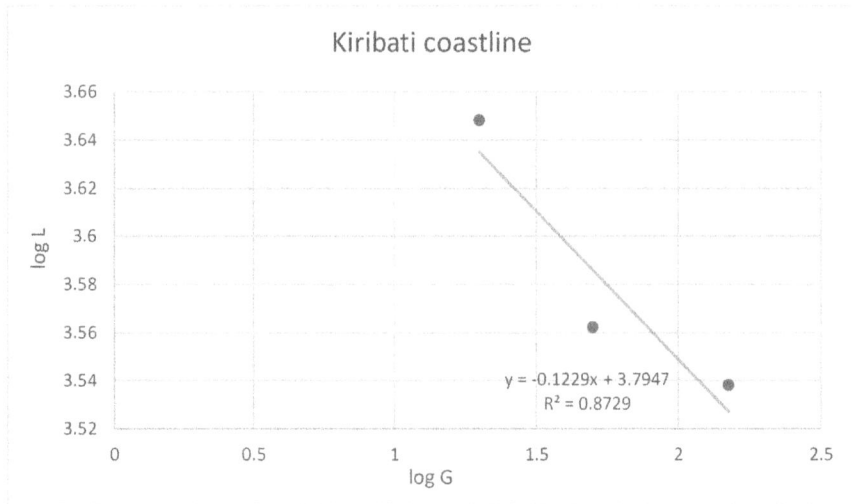

Calculating fractal dimension of Kiribati

$$D = 1 - m$$

From the graph,

$$m = -0.1229$$

$$D = 1 - (-0.1229)$$

$$D = 1.1229$$

$$D \approx 1.13$$

7. Calculating the fractal dimensions of the Bikar (Marshall islands) island

Table 2			
Length of ruler (G) (km)	50	25	12.5
log (G)	1.699	1.298	1.097
Number of rulers (n)	17	35	78
Length of coastline (km) $(L = G * n)$	850	875	962.5
log (L)	2.929	2.942	2.983

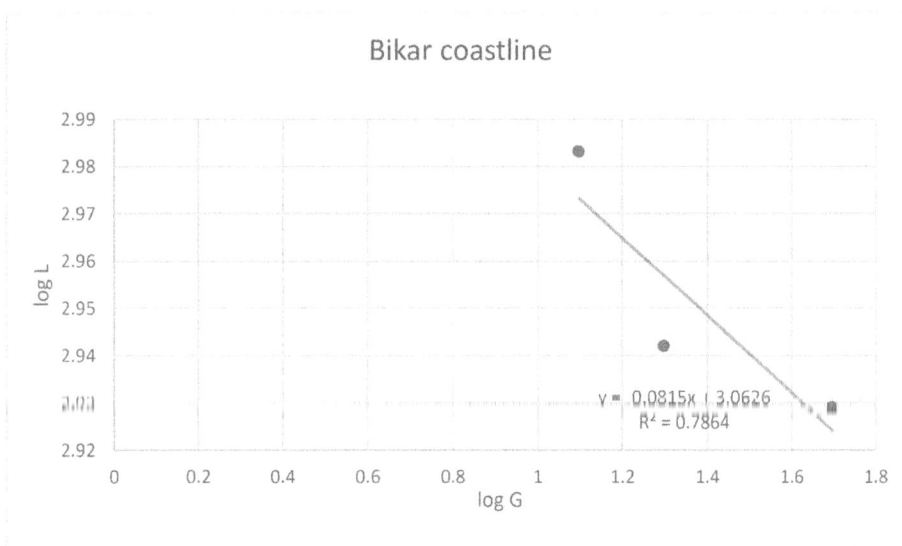

Calculating fractal dimension of Bikar

$$D = 1 - m$$

From the graph,

$$m = -0.0815$$

$$D = 1 - (-0.0815)$$

$$D = 1.0815$$

$$D \approx 1.08$$

8. Conclusion and evaluation

Consequently, following my calculations, I determined that the coastline of Kiribati and

Bikar islands have a fractal dimension of 1.13 and 1.08 respectively. The fractal

dimension I estimated for the coastline of Britain was of 3.8% accuracy.

The data shows that qualitatively Kiribati has a more complex coastline. Its coast is

more jagged than Bikar's coastline which explains the larger fractal dimension of 1.13

compared to 1.08. Nevertheless, both islands have lower fractal dimensions than Great

Britain's, which suggests that their coastlines are smoother and with less irregularity

than Britain's. It is possible to deduct that the Kiribati island has a more rugged

coastline due to increased erosion and/or the common tsunamis the islands

experiences.

As mentioned previously, the Hausdorff method is done by eye and hence is subject to a

significant margin of error. Perhaps the percentage error could be reduced with the

extension of a computer program that could precisely and accurately carry out the

placement and count of rulers on the coastlines. Another way to develop this project

would be to have other individuals to determine the fractal dimensions of the same

island using the Hausdorff method as well as the Box-counting method and contrast the results obtained. This would allow for an examination of repeated errors and a conceivable average estimation of the fractal dimensions of Kiribati and Bikar. Furthermore, it could also be expanded by conduction the Box plot method with the use of a computer program which could carry out the computations in a precise manner.

By investigating this topic, I realized how the combination of different areas of math such as regression equation and logarithmic laws, can be used to solve real-world enigmas. In addition, the concept of fractals itself was fascinating. Mandelbrot's mathematical application on real-life perplexed me and inspired me to reflect further on the subject. For example, fractal dimensions with a combination of statistical tools can be used for complex data quantification which is employed in industrial applications and assists production ([12]). This study enabled me to research a new topic and was a great foundation for estimating and describing fractal dimensions, a skill that perhaps I could apply later in my academic pursuit.

9. Bibliography

(1) RealLifeLore. (2018, March 3). The Coastline Paradox Explained [Video]. YouTube. https://www.youtube.com/watch?v=kFjq8PX6F7I

(2) Mandelbrot, B. (1967). How Long Is the Coast of Britain? Statistical Self-Similarity and Fractional Dimension. Science, 156(3775), 636-638. Retrieved November 6, 2020, from http://www.jstor.org/stable/1721427

(3) Google Maps. (n.d.). Google Maps. Retrieved 25 October 2020, from https://goo.gl/maps/evcApeK63zcK9ECU9

---

[12] *Hotař, V. (2017). Application of Fractal Dimension in Industry Practice. IntechOpen. https://doi.org/10.5772/intechopen.68276*

5. USING MODELS TO EVALUATE THE INDIAN PREMIERE LEAGUE AND PREDICT POSSIBLE FUTURE OUTCOMES OF MATCHES

Author: Vedant Sanghvi
Moderated Mark: 18/20
Level: Math AA SL

Introduction

The Indian Premiere League is the most viewed version of Twenty20 cricket throughout the world. 8 teams, representing 8 Indian cities contest in a league to attain the championship trophy. In 2014, this mammoth event was valued at over $7.2 Billion, making it one of the most valuable franchises worldwide. This sport is not only an attraction for fans and talent from all around the world, it is also the largest contributor to the betting markets of India[1]. Over the years that I've lived in India, I have become used to having an "Internal Betting War" with my friends for every single match, with one of us going back home with a treasure chest of chocolates and candies. Many a times, I won , however, the losses motivated me to explore an idea that would allow me to get that prize every time; a method for me to boost my chances of actually predicting a win. For this purpose, I have decided to attempt to model a cricket match using the statistics available for previous seasons and matches. This, paired with the correct modelling technique, will allow me to hopefully predict the outcome of any given match, and hence, will allow me to make accurate "bets". Living in Mumbai, the *Mumbai Indians* (MI) automatically become my home team. They have frequently been leaders and champions in the league.

A cricket match has 2 crucial aspects to it, first, the pitch where the match is being held. In accordance with the rules of the IPL, each team is supposed to face the other 2 times, once on each team's home ground. Second, a toss decides which team bats first (attacks) and which has the second inning to bat in (defend).

In this exploration, I will be using 2 models to predict the outcome of a match. The first model I will be using is the Poisson Distribution Model, through which I will

---

[1] "Inside Edge", Anshuman 2017

attempt to predict the number of runs each team can score in each of the 20 overs. Using this data, I will then be able to compare it against their opponents, and will be able to predict the outcome of the match to follow. The second model I will implement is a qualitative model that I have created myself, one that factors in certain variables and provides us with a viable range of scores that we can expect from each team, allowing us to once again, predict the outcome of a match.

Once I have computed the results from both models, I will compare the results and evaluate whether the model I have created is a fair and accurate model that can be applied to real life matches.

Model 1: Poisson Distribution Model

To compute for both models being used, data from the 2019 IPL league was obtained from the official IPL website[2]. The data will be used to calculate the following values which will then be implemented in the models for each match:

1. Bowling and Batting Points for team playing at Home
2. Bowling and Batting Points for team playing Away from Home

The following tables 1, 2 and 3 depict a summary of the IPL 2019 stats for each team.

---

[2] https://www.iplt20.com/points-table/2019

Team Name	Match Points			Match Statistics			
	Matches Played	Matches Won	Points	Total Runs Scored	Average Runs Scored	Total Runs Conceded	Average Runs Conceded
Mumbai Indians (MI)	14	9	18	2606	186.1428571	2471	176.5
Chennai Super Kings (CSK)	14	9	18	2473	176.6428571	2440	174.2857143
Kings XI Punjab (KXIP)	14	6	12	2429	173.5	2503	178.7857143
Kolkata Knight Riders (KKR)	14	6	12	2473	176.6428571	2429	173.5
Rajasthan Royals (RR)	14	5	11	2153	153.7857143	2129	152.0714286
Sun Risers Hyderabad (SRH)	14	6	12	2462	175.8571429	2374	169.5714286
Royal Challengers Bangalore (RCB)	14	5	11	2146	153.2857143	2266	161.8571429
Delhi Capitals (DC)	14	9	18	2546	181.8571429	2558	182.7142857
SUM	112	55	112	19288	1377.714286	19170	1369.285714
AVERAGES	14	6.875	14	2411	172.2142857	2396.25	171.1607143

Table 1 shows all the matches and the combined stats for the IPL 2019 season [3]

Games Played at Home							
Team Name	Match Points			Match Statistics			
	Matches Played	Matches Won	Points	Total Runs Scored	Average Runs Scored	Total Runs Conceded	Average Runs Conceded
Mumbai Indians (MI)	7	5	11	1208	172.5714286	1205	172.1428571
Chennai Super Kings (CSK)	7	5	11	981	140.1428571	912	130.2857143
Kings XI Punjab (KXIP)	7	5	11	1205	172.1428571	1177	168.1428571
Kolkata Knight Riders (KKR)	7	3	6	1350	192.8571429	1301	185.8571429
Rajasthan Royals (RR)	7	3	6	1138	162.5714286	1151	164.4285714
Sun Risers Hyderabad (SRH)	7	5	11	1158	165.4285714	1060	151.4285714
Royal Challengers Bangalore (RCB)	7	3	6	1076	153.7142857	1065	152.1428571
Delhi Capitals (DC)	7	4	8	1073	153.2857143	1090	155.7142857
SUM	56	33	70	19288	1312.714286	8961	1280.142857
AVERAGES	7	4.125	8.75	2411	164.0892857	2396.25	160.0178571

Table 2 shows the stats for each team when the played games on their Home Ground in the IPL 2019 season

[3] All Data Obtained is from the official IPLT20 website here.

Games Played Away from Home						
	Match Points			Match Statistics		
Team Name	Matches Played	Matches Won	Points	Total Runs Scored	Average Runs Scored	Total Runs Conceded	Average Runs Conceded
Mumbai Indians (MI)	7	5	11	1117	159.5714286	987	141
Chennai Super Kings (CSK)	7	5	11	1184	169.1428571	1247	178.1428571
Kings XI Punjab (KXIP)	7	5	11	1224	174.8571429	1326	189.4285714
Kolkata Knight Riders (KKR)	7	3	6	1123	160.4285714	1128	161.1428571
Rajasthan Royals (RR)	7	3	6	1015	145	1041	148.7142857
Sun Risers Hyderabad (SRH)	7	5	11	1204	172	1214	173.4285714
Royal Challengers Bangalore (RCB)	7	3	6	1070	152.8571429	1201	171.5714286
Delhi Capitals (DC)	7	4	8	1473	163.6666667	1468	163.1111111
SUM	56	33	70	19288	1297.52381	9612	1326.539683
AVERAGES	7	4.125	8.75	2411	162.1904762	2396.25	165.8174603

Table 3 shows the stats for each team when they played games Away from their Home Ground in the IPL 2019 season

To attain values for use in the Poisson model, we will use the data above to calculate the bowling & batting strength for each team, Home and Away.

Each Strength is further subdivided for each team, wherein, we calculate their strength when playing at home, and when playing away. The following method is used to obtain the values:

1. Calculating the average runs scored when playing at home and when playing away.

   This is done using the simple formula for obtaining mean ($\frac{\Sigma n}{N}$), wherein "n" is the number of runs scored in each game and "N" is the total number of games played. Considering the Mumbai Indians, we use the following calculations to attain the values:

a) Home Batting Average

   1208/7 = 172.57 runs scored per match at home

b) Away Batting Average

   1117/7 = 159.57 runs scored per match away from home

Once we have similar values for each team, we compare them to the cumulative average of Home batting (164.08) and Away Batting (162.19). A team's Batting score is then attained by obtaining the difference between each team's individual score and the mean scores.

2. Using the same method, we obtain the bowling scores for each team when playing at home and when playing matches away; however, we will consider the runs conceded in each match. For example, for the Mumbai Indians,

   a) Home Bowling Average

      1205/7 = 172.14 runs conceded per match at home

   b) Away Bowling Average

      987/7 = 141 runs conceded per match away from home

To observe the practical implementation of each of these scores, we must take an example of a match. Say the Mumbai Indians are playing the Chennai Super Kings at the Wankhede Stadium (Mumbai Indians' Home Ground).

In this case, we know that the Mumbai Indians are playing at their home ground, however, we do not know who will be attacking or defending, since the toss is not yet conducted. Hence, we will look to calculate the Score Rating (SR) for each team. This will be done using the following steps:

1. For Mumbai Indians:

   a) Divide the Average runs scored at home by the Mumbai Indians by the cumulative average runs scored by all teams in home games

   $$\therefore, 172.57 / 164.09 = 1.0516$$

   The value obtained shows us that the Mumbai Indians score approximately 5% more runs than the league average when playing at home.

   b) Divide the Average runs conceded at home by the Mumbai Indians by the cumulative average runs conceded by all teams in home games

   $$\therefore, 172.14 / 160.01 = 1.0758$$

   The value obtained shows us that the Mumbai Indians concede approximately 7.5% more runs than the league average when playing at home.

2. For Chennai Super Kings:

   a) Divide the Average runs scored away from home by the Chennai Super Kings by the cumulative average runs scored by all teams in away games

   $$\therefore, 169.14 / 162.19 = 1.0428$$

   The value obtained shows us that the Chennai Super Kings score approximately 4.3% more runs than the league average when playing at home.

   c) Divide the Average runs conceded at away from home by the Chennai Super Kings by the cumulative average runs conceded by all teams in away games

   $$\therefore, 178.14/165.82 = 1.0742$$

The value obtained shows us that the Chennai Super Kings concede approximately 7.4% more runs than the league average when playing away from home.

3. Calculating the Score for the Mumbai Indians:

The Mumbai Indians' score would then be calculated by using the following formula:

$$MI's\ Score = (\ MI's\ Home\ Batting\ Strength * CSK's\ Away\ Bowling\ Strength$$
$$* Average\ Runs\ Scored\ By\ MI\ in\ home\ games)/\ 20$$

$$= \frac{(1.0516 * 1.0742 * 172.57)}{20}$$

This gives us a Score of 9.75, meaning that the MI is predicted to score approximately 9.75 runs in each over of the game against the Chennai Super Kings. This would mean they put up a total of 195 runs.

4. Calculating the Score for the Chennai Super Kings:

The Mumbai Indians' score would then be calculated by using the following formula:

$$CSK's\ Score = (CSK's\ Away\ Batting\ Strength$$
$$* MI's\ Home\ Bowling\ Strength$$
$$* Average\ runs\ scored\ by\ CSK\ in\ an\ Away\ match)/20$$

$$= \frac{(1.0428 * 1.0758 * 169.14)}{20}$$

This gives us a Score of 9.49, meaning that the Chennai Super Kings can score up to 9.5 runs every over against the Mumbai Indians' bowling, meaning that they would put up a total of 190 runs.

These scores create a dubious outcome prediction. Both teams seem extremely competent and either being able to win the match with an extremely scarce number of runs to spare.

However, we must consider that most of the values used in the formulae above are rounded off to 2 decimal places, hence leaving certain room for error. To eliminate this, we now implement the Poisson model to allow us to gain a better and more precise insight into all the possibilities of the match. Once again, in our formula we will be considering 2 teams and will be exploring all possibilities of outcomes in the 20 overs of the game.

For the purpose of my exploration, I will be using an adaptation of the original Poisson model formula, as such:

$$P(H, A) = \frac{\Delta^H * e^{-\Delta}}{H!} * \frac{\Omega^A * e^{-\Omega}}{A!} * 100$$

Herein, "H" and "A" are the rounded off values of the predicted scores of each team; i.e., 9 and 9 for each team. The values are rounded down since scoring more than the 9.75 and the 9.49 value might not be possible. "Δ" and "Ω" stand for the exact predicted scores per over for the MI and CSK respectively.

Hence, each variable has the following values:

H = 9

Λ - 0

Δ = 9.747001

Ω = 9.487436

The following values are thus substituted in the formula:

$$P(9,9) = \frac{9.747001^9 * e^{-9.747001}}{9!} * \frac{9.487436^9 * e^{-9.487436}}{9!} * 100$$

$$= 0.1279402849 * 0.1300875092 * 100$$

$$= 1.664343299 \%$$

The low percentage value we obtain from the Poisson formula suggests that the probability of the Mumbai Indians and Chennai Super Kings each scoring 9 runs per over is quite low. This result now demands that we calculate the probability of scores throughout the match. In the 2018 season of the IPL, the top score attained by a team was approximately 240 runs, averaging out to 12 runs per over. Using this value, we will try to calculate the probability of each team scoring up to a maximum of 12 runs per over, using the same formula as above. The results are displayed in the table below:

		Away Team (Chennai Super Kings)												
		0	1	2	3	4	5	6	7	8	9	10	11	12
	0	4.4319E-07	4.2047E-06	1.9946E-05	6.3079E-05	0.000149615	0.000283892	0.000448901	0.000608	0.000722	0.000760618	0.000722	0.000622	0.000492
	1	4.31978E-06	4.0984E-05	0.00019441	0.00061483	0.001458295	0.002767096	0.004375441	0.00593	0.007033	0.007413749	0.007034	0.006067	0.004796
	2	2.10524E-05	0.00019973	0.00094748	0.00299638	0.007107001	0.001383548	0.021323713	0.028901	0.034275	0.036130908	0.034279	0.029565	0.023375
	3	6.83993E-05	0.00064893	0.00307836	0.00973525	0.023090648	0.043814209	0.069280751	0.0939	0.111358	0.117389332	0.111372	0.096058	0.075945
Home	4	6.83993E-05	0.00158129	0.0075012	0.02372238	0.056266143	0.106764286	0.168819888	0.22881	0.271352	0.286048483	0.271387	0.234069	0.18506
Team	5	0.000324911	0.00308257	0.01462284	0.04624441	0.10968523	0.20812632	0.329097524	0.446042	0.528974	0.557622971	0.529041	0.456295	0.360756
(Mumbai	6	0.000527817	0.00500763	0.0237548	0.07512406	0.178183675	0.338101242	0.534618982	0.724595	0.859318	0.905858609	0.859428	0.741251	0.586048
Indians)	7	0.000734948	0.00697277	0.03307687	0.1046049	0.248108065	0.470781878	0.744418822	1.008947	1.196539	1.261343539	1.196692	1.03214	0.81603
	8	0.000895443	0.00849545	0.04030004	0.127448	0.302288695	0.573588929	0.906981375	1.229275	1.457834	1.536789591	1.458019	1.257533	0.99423
	9	0.000969764	0.00920058	0.04364494	0.1380262	0.32737869	0.621196874	0.98226093	1.331305	1.578834	1.664343298	1.579035	1.361909	1.076752
	10	0.000945229	0.0089678	0.04254073	0.13453415	0.319096042	0.605480655	0.957409827	1.297623	1.53889	1.622235579	1.539086	1.327452	1.04951
	11	0.000837559	0.00794629	0.03769496	0.1192095	0.282748131	0.536510959	0.848352231	1.149812	1.363597	1.437448347	1.36377	1.176244	0.929961
	12	0.000680308	0.00645437	0.03061773	0.09682793	0.229662193	0.435781071	0.68907417	0.933935	1.107581	1.167567539	1.107722	0.955404	0.755361

Table 4 shows the Poisson Distribution results for every probability in the MI vs CSK match

The sum of all the cells in the table comes up to 68.23554851 points. Herein, the yellow cells represent situations of a tie, wherein both teams score exactly the same number of runs. This adds up to 8.141550142 points, meaning that there is approximately a 11.9% chance that the match may end up in a tied situation. Any values above the yellow markers represent a win for the Chennai Super Kings. This section adds up to 28.72469709 points, suggesting that the Chennai Super Kings have an approximated 42.1% chance of securing a win in the match. Finally, the section for the Mumbai Indians (anything below the yellow marker) adds up to a value of 31.36930127 points, signifying that they have a greater chance of securing a win, at almost 46%.

Using these statistics, we can thus safely predict that the Mumbai Indians have an almost definitive chance of securing a win against the Chennai Super Kings. This is specifically given the circumstances of Mumbai playing at the *Wankhede* Stadium.

Model 2: The League Score Evaluator (LSE) Model

The LSE model is one that I have attempted to develop myself. To an extent, the theory is based upon calculations that might be derived from the Poisson distribution model, however, the procedure applied is different.

For this model, we will be using the data from Tables 1, 2 and 3 above to calculate two aspects of each team, their Home Rating and their Away Rating. These will then be applied to a simple mathematical procedure to obtain a prediction.

The Batting and Bowling Ratings are further divided into 2 sub-categories, the attack and defence ratings. Once again, the team "attacking" is the one that bats first, while the team "defending" is the one that bats in the second innings of the match.

The following equations will be used to derive the value of each of these factors:

1. Batting Rating

   a) Home Attack Rating (HAR)

$$HAR = \frac{Number\ of\ Runs\ scored\ by\ team\ in\ an\ Attacking\ Situation\ at\ Home}{Total\ runs\ scored\ by\ all\ teams\ in\ an\ Attacking\ Situation\ at\ Home}$$

   b) Home Defence Rating (HDR)

$$HDR = \frac{Number\ of\ Runs\ scored\ by\ team\ in\ a\ Defending\ Situation\ at\ Home}{Total\ runs\ scored\ by\ all\ teams\ in\ a\ Defending\ Situation\ at\ Home}$$

   c) Away Attack Rating (AAR)

$$AAR = \frac{Number\ of\ Runs\ scored\ by\ team\ in\ an\ Attacking\ Situation\ Away\ from\ Home}{Total\ runs\ scored\ by\ all\ teams\ in\ an\ Attacking\ Situation\ Away\ from\ Home}$$

   d) Away Defence Rating (ADR)

$$ADR = \frac{Number\ of\ Runs\ scored\ by\ team\ in\ a\ Defending\ Situation\ Away\ from\ Home}{Total\ runs\ scored\ by\ all\ teams\ in\ a\ Defending\ Situation\ Away\ from\ Home}$$

Looking at the scores of the Mumbai Indians, as shown in Table 5 below, they have scored 528 runs in a home attack situation. The overall home attack runs in the IPL 2019 are 5471. Thus, using for formula for the HAR, we obtain Mumbai Indians' HAR value as 0.0965. Similarly, the Table 6 below shows each corresponding value for every team in the league.

| Team | Runs Scored | | | |
| | Home Games | | Away Games | |
	Attack Games	Defence Games	Attack Games	Defence Games
Mumbai Indians	528	680	1132	266
Chennai Super Kings	821	291	457	904
Royal Challengers Bangalore	717	359	612	458
Kolkata Knight Riders	964	286	400	723
Delhi Capitals	658	415	684	789
Sun Risers Hyderabad	447	711	1003	301
Kings XI Punjab	855	350	544	680
Rajasthan Royals	481	657	483	532

Table 5 shows the parametric runs scored for each component for each team

Teams	Scoring Parameters			
	HAR	HDR	AAR	ADR
Mumbai Indians	0.0965	0.1579	0.2129	0.0663
Chennai Super Kings	0.1501	0.1513	0.0854	0.259
Royal Challengers Bangalore	0.1311	0.0987	0.1151	0.0663
Kolkata Knight Riders	0.1762	0.0461	0.0753	0.1145
Delhi Capitals	0.1203	0.1382	0.1287	0.1988
Sunrisers Hyderabad	0.0817	0.1908	0.1887	0.0783
Kings XI Punjab	0.1563	0.0789	0.1024	0.1205
Rajasthan Royals	0.0879	0.1382	0.0909	0.0964

Table 6 shows the calculated HAR, HDR, AAR & ADR values for each team

The next step in the process is to obtain the average runs that each team scored in all these situations. Once again, looking at the Mumbai Indians, they have scored 528 runs in their 3 home attack games. This brings their average score to 176 runs per home attack game. Similarly, the set of tables below show the corresponding values for each team.

Using Models to Evaluate the Indian Premiere League and Predict Possible Future Outcomes of Matches

Mumbai Indians' Average Runs	Home Games	Away Games
Attack	176	161.7
Defence	170	133

Chennai Super Kings' Average Runs	Home Games	Away Games
Attack	164.2	152.3
Defence	97	150.6

Royal Challengers Bangalore's Average Runs	Home Games	Away Games
Attack	179.25	153
Defence	179.5	152.6

Delhi Capitals' Average Runs	Home Games	Away Games
Attack	164.5	171
Defence	138.3	157.8

Kolkata Knight Riders' Average Runs	Home Games	Away Games
Attack	192.8	133.3
Defence	193	180.75

Sunrisers Hyderabad's Average Runs	Home Games	Away Games
Attack	223.5	167.2
Defence	142.2	150.5

Kings XI Punjab's Average Runs	Home Games	Away Games
Attack	171	178
Defence	175	170

Rajasthan Royals' Average Runs	Home Games	Away Games
Attack	160.3	161
Defence	164.25	177.3

Based on the results as above, we can use the following method to evaluate the progress and outcome of a match:

1.   Say the Mumbai Indians are playing the Chennai Super Kings at the Wankhede Stadium (MI's home-ground.)

2.   The Mumbai Indians win the toss and elect to bat first, meaning that they would be the ones attacking.

3.   We would then multiply MI's HAR with the average score they make in every home match; 0.0965 * 176 = 16.984 runs. According to my theory and model, The Mumbai Indians would have a predicted score within ±17 runs of their average 176 score, i.e, between 159 to 193.

5.   We would then have to compare CSK's average away defence runs to CSK's ADR; 0.259 * 152.3 = 39.4457, meaning that they have a boundary of appx. 39 runs to match the target that The Mumbai Indians set.

6.   If the Mumbai Indians reach their predicted top score of 193, the CSK would just fall short of reaching the target, meaning that they would need an external factor to benefit their side.

Looking at the conclusion I derived from using this model, I can fairly say that it supports the conclusion we derived from the Poisson model as well. The results there showed a thin margin for either of the two teams to win.

Discussion and Evaluation

Cricket is an extremely qualitative game. There are factors that models cannot combine and assess. Thus, there are certain limitations and certain assumptions that we must consider and maintain when using this model.

1. Any IPL team is an extremely dynamic family. Players are dropped and picked up in the middle of an existing season; players are injured, sometimes are benched for a long time. In addition to this, there are times when players just don't perform to expectations.

2. Each pitch and each ground develops over the course of a season; one pitch may be dry; one may have a little extra bounce or some may be more inclined towards a spin.

3. Inclement weather is an obstruction to many matches. India is known for its sometimes-torrential rains that might lead to matches being delayed and cancelled. Delayed matches often affect the morale of players and this can be a factor affecting a match's outcome.

4. Fatigue and injuries are further unavoidable circumstances that chance the dynamic of a team and its performance in a match.

For this model to be used, the following assumptions must be kept in mind:

1. Throughout the season, the playing squad must remain the same, no substitutes and no dropouts from the team during matches.
2. All players are playing at their optimal level, assuming that they are under no emotional/ psychological duress and are at the best of their health.
3. There is no alteration in pitch conditions. As far as this model is concerned, the players are playing on the same pitch conditions, with the surrounding weather remaining constant too.

This exploration focuses on the use of models to evaluate a real-life situation and game. The use of mathematic and statistical models allows us to quantify situations and understand the processes that lead to an outcome. They further allow us to alter certain variables and investigate their effect on the system without having to actually make the changes in real-life. Finally, models allow users to make prediction in regards to the outcome of changing factors before they occur in real-life; helping designers prevent flaws and disruptions. On the other hand, models cannot in cooperate all the various aspects of a real-life event or situation. Not all qualitative variables can be quantified and hence cannot be added to a set of formulae or functions. This raises questions as to the predictions and the conclusions drawn from mathematical models. The Poisson model that I have applied here would prove to be extremely useful, given that the user was adept at math and could make quick mental calculations. However, if a user was a student like myself, or even a younger person, they might find it a little challenging to process all the calculations quick enough to place their bets. It would be necessary to develop a table like Table 4 for each and every match possible; this would have to be done before a particular match starts. In this view, the LSE model that I suggested helps in cutting short the time required to make the calculation, while perhaps issuing results in the same boundary of error as the Poisson model does.

It is however important for us to acknowledge that the Poisson distribution model is one that is recognized around the world and has been proven to be effective countless times. On the other hand, my LSE model is one that does not have the support of academia and has not been recognized outside my sphere of influence. The LSE model might be useful for young adults such as myself who are interested in play-betting on cricket matches, however, cannot be used at a professional level.

6. MODELLING THE PHARMACOKINETIC PROFILE OF ERYTHROMYCIN

Author: David Sacoor
Moderated Mark: 18/20
Level: Math AI HL

Modelling the pharmacokinetic profile of erythromycin

Introduction

A few months ago, I studied about natural selection and antibiotic resistance in IB Biology class. The rates of antibiotic resistance have increased significantly worldwide over the past few years. Some of the causes of this problem include over-prescription of antibiotics by doctors and patients not finishing the entire antibiotic course. However, the main cause of antibiotic resistance is the overuse of antibiotics in fish farming and intensive animal agriculture. This is a topic of personal interest because I want to be a doctor and antibiotic biological behavior and resistance are topics of major importance when doctors have to prescribe antibiotics to patients who need them. Antibiotics are crucial tools that doctors have to combat bacterial infections in patients. One of my family members had pneumonia in the past and regularly took an antibiotic called erythromycin. Subsequently, I was inclined to gain a better understanding of the pharmacokinetics of erythromycin as this was required by the doctor who designed a suitable erythromycin regimen. Pharmacokinetics is the study of how drugs are absorbed into, distributed and metabolised into, and excreted from, the body.[1]

The aim of this exploration is to find adequate nonlinear regressions which model the serum concentration of erythromycin in relation to time. In order to find these mathematical models, I will use two different methods (polynomial interpolation and automated nonlinear regression) and then compare them. Further, I will then investigate these models using differential and integral calculus in order to arrive at two pharmacokinetic profiles. This will enable me to find the C_{max}, the t_{max} and the

---

[1] https://medical-dictionary.thefreedictionary.com/. 2004. *Pharmacokinetics*. [online] Available at: <https://medical-dictionary.thefreedictionary.com/pharmacokinetics> [Accessed 1 December 2020].

AUC of the pharmacokinetic profile of erythromycin. The definition of these key terms can be found in the table below. Erythromycin can be administered in many ways, such as orally, intravenously, and intramuscularly. My family member was administered the oral form of erythromycin. The graph below illustrates what these

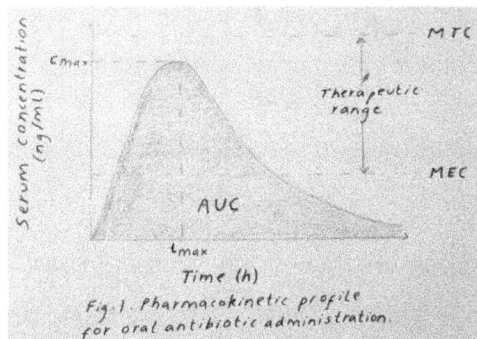

pharmacokinetic symbols mean in a pharmacokinetic profile. The C_{max} and AUC are particularly important pharmacokinetic exposure parameters because they are used to assess the toxicity and efficacy of a certain antibiotic. After being administered, erythromycin is absorbed in the epithelial lining of the small intestine, then distributed to most body tissues, next it is metabolised by the liver and finally it is excreted in the bile.

Key term	Definition
C (ng/ml)	Serum concentration in nanograms per milliliter (ng/ml).
t (h)	Time in hours (h).
C_{max} (ng/ml)	Maximum (peak) serum concentration after drug administration.
t_{max} (h)	Time to reach maximum (peak) serum concentration following drug administration.
AUC (h ng ml^{-1})	Area under the serum concentration-time curve, which represents the total body exposure to the drug.
MEC	Minimum effective serum concentration required to achieve the desired pharmacological response.
MTC	Minimum toxic concentration.
$Therapeutic\ range$	Range of serum concentration required to achieve desired therapeutic effect.

Table 1-Table with definition of key pharmacokinetic terms in the exploration[2]

[2] Anon, Clinical Pharmacokinetics Preferred Symbols. *link.springer.com.* Available at: https://static.springer.com/sgw/documents/1372030/application/pdf/40262_cpk_symbols.pdf [Accessed August 17, 2020].

In order to develop both of my models, I firstly collected data from *Symbiosis Online Publishing*, more specifically from a research article entitled "Mathematical Model of the Pharmacokinetic Behavior of Orally Administered Erythromycin to Healthy Adult Male Volunteers (Ďurišová)[3]."

Time (h)	0	0.5	1.0	2.0	4.0	6.0	9.0
Serum concentration of erythromycin (ng/ml)	0.0	47.4	185.0	293.0	254.0	158.0	76.9

Table 2-Table to show the serum concentration of erythromycin over time.

Since both the time and the serum concentration of erythromycin are always ≥ 0, the graph will be plotted in the first quadrant. I used the modelling software SciDAVis to plot a scatter graph with time (h) on the $x-$axis and the erythromycin serum concentration (ng/ml) on the $y-$axis.

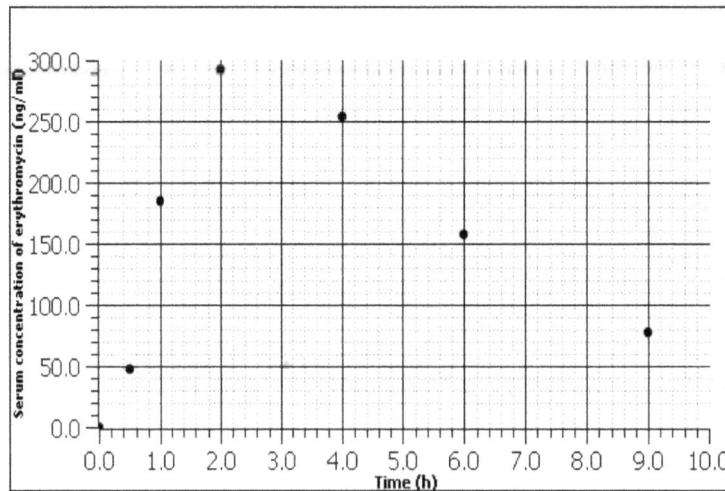

Fig. 2. Scatter graph of time (h) vs Erythromycin serum concentration (ng/ml)

[3] Ďurišová, M., 2015. *Mathematical Model Of The Pharmacokinetic Behavior Of Orally Administered Erythromycin To Healthy Adult Male Volunteers.* [online] Symbiosisonlinepublishing.com. Available at: <https://symbiosisonlinepublishing.com/pharmacy-pharmaceuticalsciences/pharmacy-pharmaceuticalsciences25.pdf> [Accessed 17 August 2020].

144

Firstly, by visual inspection of the data points, it is unmistakable that nonlinear regression is more appropriate to model the data than linear regression. By analysis of the scatter graph, we can see that there is a maximum point of 293 ng/ml at 2 hours. Hence, $C_{max} = 293$ ng/ml and $t_{max} = 2$ h.

Before the modelling process, I will analyse the data in order to facilitate a better understanding of the relationship between the two variables. First of all, I observe in the data that from $t = 0$ to $t = 2$, the serum concentration of erythromycin rises rapidly to peak (at the C_{max}). To illustrate this, I calculated the average rate of change between each of the data points; this is shown below. A positive and increasing first rate of change up to $t = 1$ means that the model function must be increasing up to this point. From $t = 0$ to $t = 1$, $\frac{\Delta^2 C}{\Delta t^2}$ is > 0 which suggests the graph is concave upwards during this period. At $t = 2$, $\frac{\Delta^2 C}{\Delta t^2}$ becomes negative, which implies that there is a point of inflexion between $t = 1$ and $t = 2$, when $\frac{\Delta^2 C}{\Delta t^2} = 0$. From $t = 2$ to $t = 6$, $\frac{\Delta^2 C}{\Delta t^2}$ is < 0, indicating that the graph is concave downwards. From $t = 2$ to $t = 9$, the average rate of change is negative, which means the model function must be decreasing during this period. Withal, from $t = 4$ to $t = 9$, the average rate of change is negative and initially increases, but then it decreases. Further, the concavity of the graph changes again between $t = 6$ and $t = 9$, which hints at the presence of a point of inflexion between this period. It goes without saying that an ideal model function should pass through the point $(0,0)$.

Method A: Using polynomial interpolation

Interpolation is the estimation of the value of a function $f(x)$ from certain known values of the function. Given n distinct points (x_i, y_i) with $i \in \{1, 2, \ldots, n\}$, we call polynomial interpolation when determining a polynomial $P(x)$ of degree $\leq n - 1$ such that $P(x_i) = y_i$ for every i.

There are $C_4^7 = 35$ possible combinations of four points out of seven. This would lead to 35 systems of cubic equations, with each one yielding a different polynomial with a different goodness of fit. As it would be very time-consuming to solve these 35 equations and then find the one which gives the best fit, another possible way to determine this polynomial without having to do so is to use the Lagrange interpolation formula, which is simpler and more efficient. As stated by Arashiro et al., the Lagrange interpolation formula is a way to find a polynomial which takes on certain values (of the independent variable) at arbitrary points.[5] The formula is shown below:

$$P(x) = \sum_{i=1}^{n} L_i(x) y_i$$

where $1 \leq i \leq n$ and $P(x_i) = y_i$, and where n is the unique polynomial of degree $n - 1$ that approximates the function $y = P(x)$ given n data points as $(x_1, y_1), (x_2, y_2), \ldots, (x_{n-1}, y_{n-1}), (x_n, y_n)$.

$L_i(x), i \in \{1, 2, \ldots, n\}$, are n Lagrange polynomials of degree $n - 1$ defined[6] as

$$L_i(x) = \prod_{j=1, j \neq i}^{n} \frac{x - x_j}{x_i - x_j} = \frac{(x - x_1)\ldots(x - x_{i-1})(x - x_{i+1})\ldots(x - x_n)}{(x_i - x_1)\ldots(x_i - x_{i-1})(x_i - x_{i+1})\ldots(x_i - x_n)}$$

[5] Arashiro, T., n.d. *Lagrange Interpolation | Brilliant Math & Science Wiki.* [online] Brilliant.org. Available at: <https://brilliant.org/wiki/lagrange-interpolation/#sample-problems> [Accessed 20 August 2020].
[6] Tatum, J., 2020. *1.11: Fitting A Polynomial To A Set Of Points - Lagrange Polynomials And Lagrange Interpolation.* [online] Physics LibreTexts. Available at: <https://phys.libretexts.org/Bookshelves/Astronomy__Cosmology/Book%3A_Celestial_Mechanics_(Tatum)/01%3A_Numerical_Methods/1.11%3A_Fitting_a_Polynomial_to_a_Set_of_Points_-_Lagrange_Polynomials_and_Lagrange_Interpolation> [Accessed 14 November 2020].

Time (h)	Serum concentration of erythromycin (ng/ml)	$\frac{\Delta C}{\Delta t}$	$\frac{\Delta^2 C}{\Delta t^2}$
0.0	0.0	0.0	0.0
0.5	47.4	94.8	189.6
1.0	185.0	275.2	360.8
2.0	293.0	108.0	-167.2
4.0	254.0	-19.5	-63.8
6.0	158.0	-48.0	-14.3
9.0	76.9	-27.0	7.0

Table 3-First and second average rates of change between the data points.

Looking at the distribution of the data points in *Figure 1*, one type of function which may model the points is a cubic polynomial. This is because the data points suggest that there is one real zero. A quadratic polynomial function is not appropriate to model my data because the distribution of data points is not symmetric and also because there are two points of inflexion in my graph. I then used my GDC to evaluate the goodness of fit of a cubic polynomial for my data, and it yielded a high coefficient of determination (R^2) of 0.954. The coefficient of determination is a measure that provides information about the goodness of fit of a model. In the context of regression, it is a statistical measure of how well the regression line approximates the actual data (Newcastle University)[4]. A cubic model would have a point of inflexion and, looking at the plot in graph 1, there does in fact appear to exist a change in concavity between $t = 6$ and $t = 9$, when $\frac{\Delta^2 C}{\Delta t^2}$ changes sign from negative to positive.

[4] https://internal.ncl.ac.uk/. 2018. *Coefficient Of Determination, R-Squared*. [online] Available at:
<https://internal.ncl.ac.uk/ask/numeracy-maths-statistics/statistics/regression-and-correlation/coefficient-of-determination-r-squared.html> [Accessed 14 November 2020].

In order to find a cubic polynomial model, I need 4 data points. The four data points I chose are shown below. These were chosen as they are equally distributed among the seven data points.

i	t	C
1	0	0
2	1.0	185.0
3	4.0	254.0
4	9.0	76.9

Table 4- t and C coordinates of the four data points chosen.

$$P(t) = \sum_{i=1}^{4} L_i(t)C_i$$

$$= L_1(t)C_1 + L_2(t)C_2 + L_3(t)C_3 + L_4(t)C_4$$

Recall that

$$L_i(x) = \prod_{j=1, j \neq i}^{n} \frac{x - x_j}{x_i - x_j}$$

Hence, the four Lagrange polynomials are

$$L_1(t) = \left(\frac{t-1}{0-1}\right)\left(\frac{t-4}{0-4}\right)\left(\frac{t-9}{0-9}\right)$$

$$L_2(t) = \left(\frac{t-0}{1-0}\right)\left(\frac{t-4}{1-4}\right)\left(\frac{t-9}{1-9}\right)$$

$$L_3(t) = \left(\frac{t-0}{4-0}\right)\left(\frac{t-1}{4-1}\right)\left(\frac{t-9}{4-9}\right)$$

$$L_4(t) = \left(\frac{t-0}{9-0}\right)\left(\frac{t-1}{9-1}\right)\left(\frac{t-4}{9-4}\right)$$

Therefore, the polynomial of degree 3 that passes through the 4 data points chosen is, then,

$$P(t) = \frac{(t-1)(t-4)(t-9)}{(0-1)(0-4)(0-9)} \times 0 + \frac{(t-0)(t-4)(t-9)}{(1-0)(1-4)(1-9)} \times 185$$

$$+ \frac{(t-0)(t-1)(t-9)}{(4-0)(4-1)(4-9)} \times 254 + \frac{(t-0)(t-1)(t-4)}{(9-0)(9-1)(9-4)} \times 76.9$$

Although I could have expanded the separate algebraic expressions and summed them, this would have been monotonous and senseless. On that account, I decided that the best method to find the cubic polynomial model was to use an online Lagrange interpolation calculator[7]. The equation will be given precise to 3 decimal places because this is the maximum number of decimal places provided by this calculator. Thus, the equation of my cubic polynomial model is given by:

$$P(t) \approx 3.689t^3 - 58.943t^2 + 240.254t + 0$$

I then graphed this model function on top of the data points using SciDAVis.

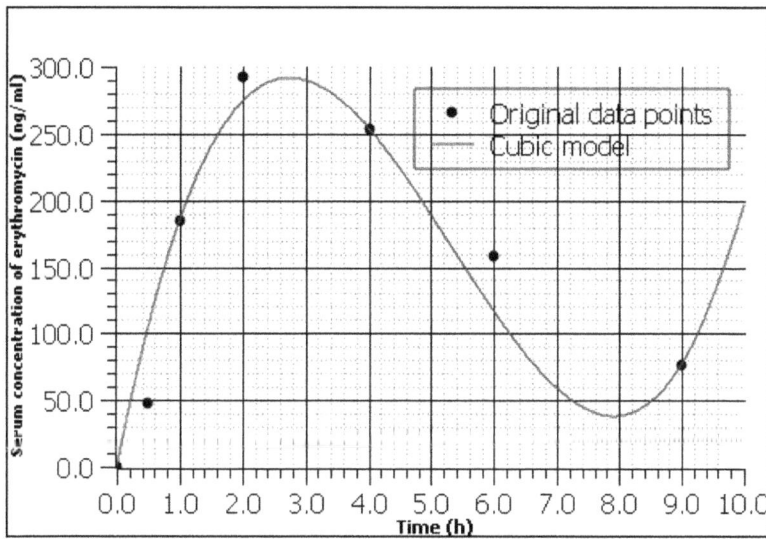

Fig. 3. Cubic model (blue) compared to the original data points (black).

[7] Dcode.fr. 2020. *Lagrange Interpolation Polynomial Calculator - Online Tool*. [online] Available at: <https://www.dcode.fr/lagrange-interpolating-polynomial> [Accessed 22 August 2020].

149

As shown, my cubic polynomial models the erythromycin serum concentration quite closely from 0 to around 7.5 hours, in the sense that it resembles the shape of a pharmacokinetic profile within this particular domain. In addition, a strength of my cubic model is that it has a C-intercept of 0. However, after this time, my model is incorrect because as $t \to \infty$, the erythromycin serum concentration also $\to \infty$. This is wrong based on the basic pharmacokinetic profile shown on page 2 (*Figure 1*). In a pharmacokinetic profile, the serum concentration of drug reaches a maximum (the C_{max}) and then decreases as $t \to \infty$ (as shown in *Figure 1*). Moreover, after looking at *Figure 1*, as $t \to \infty$, the serum concentration approaches the line $t = 0$ but never touches or crosses it. Thus, the time axis is a horizontal asymptote in a pharmacokinetic profile. Lastly, the reason why it does not make biological sense for the erythromycin serum concentration to $\to \infty$ as $t \to \infty$, is because after the antibiotic is absorbed into the bloodstream, it is then metabolised and excreted (the two routes of drug elimination). This causes its concentration to decrease. While this model function passes precisely through four of data points, the presence of two turning points ultimately make it unsuitable to represent how the concentration of a drug in the serum changes over time. For all these reasons, my cubic model clearly cannot be extrapolated for $t > 7.5$.

Nonetheless, I will now analyse the cubic polynomial model I created with differential calculus and integral calculus, in order to find the key pharmacokinetic exposure parameters (C_{max}, t_{max} and AUC) of erythromycin.

The purpose of differentiating my cubic model is to estimate the time, in hours, (the t_{max}) when the serum concentration of erythromycin (ng/ml) reaches its maximum value (the C_{max}). To find the first derivative of my model function, I need to apply the power rule.

Hence,

$$\frac{d}{dt}(3.689t^3 - 58.943t^2 + 240.254t + 0) = 11.067t^2 - 117.886t + 240.254$$

Using the first derivative, I can now compare the C_{max} of my cubic model to that of the data.

At the local maximum (C_{max}), $P'(t) = 0$

Thus,

$$11.067t^2 - 117.886t + 240.254 = 0.$$

To solve this quadratic equation, I used my GDC to graph it and find the points of intersection with the time axis.

$$t \approx 2.746 \text{ and } t \approx 7.906$$

The t_{max} for the data is 2. Therefore, I will use the value of t of 2.746 and reject the other one.

Now, I can find the C_{max} of my cubic polynomial model using this value of t.

$$C_{max} = P(2.746)$$

$$= 3.689(2.746)^3 - 58.943(2.746)^2 + 240.254(2.746) + 0$$

$$\approx 291.7 \text{ ng/ml (precise to 1 decimal place)}$$

The C_{max} of my model function is close to the C_{max} of the data, which is 293.0 ng/ml. This is a clear strength of my cubic model.

In order to find the bioavailability of erythromycin, we need to calculate the AUC of the cubic polynomial model. To find the AUC, we need to find the definite integral between 0 (the lower limit of the integral) and 9 (the upper limit of the integral) hours.

Consequently,

$$AUC = \int_0^9 (3.689t^3 - 58.943t^2 + 240.254t + 0)dt$$

To integrate my cubic polynomial model, I need to use the power rule. So, the AUC is equal to

$$\int_0^9 (3.689t^3 - 58.943t^2 + 240.254t + 0) \, dt$$

Applying the rules of definite integration:

$$AUC = \left[\frac{3.689t^4}{4} - \frac{58.943t^3}{3} + \frac{240.254t}{2} + 0t \right]_0^9$$

I then used a GDC to evaluate AUC. This gives me:

$$AUC \approx 1458 \, \text{h ng ml}^{-1} \text{ (rounded to the nearest whole integer)}$$

We can show the AUC graphically using SciDAVis.

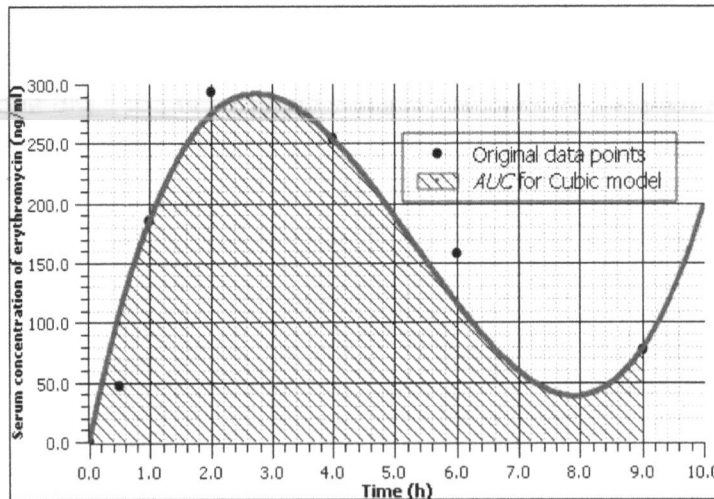

Fig. 4. Graphical representation of the AUC for my cubic model.

Method B: Automated nonlinear regression

One of the limitations of my cubic model is that the time axis is not a horizontal asymptote. Subsequently, I will now use technology (namely a programme called

Curve Expert[8]) to find a function which models my data more accurately. It is important to state that even though my domain is restricted to $\{0 < t < 9\}$, the model must evidently allow for realistic nonlinear extrapolation (which my cubic model clearly does not). This programme uses nonlinear least squares fittings to find functions that best fit a given set of data points; results are shown from the best suiting function (highest R^2 value) to the worst suiting function (lowest R^2 value).

The function which best models my data is called a Hoerl function. Crucially, besides the fact that it yields an $R^2 \approx 0.987$, it has a horizontal asymptote of $C = 0$. This is what makes it a more suitable model over the cubic model: it accounts for the plateauing in erythromycin serum concentration as $t \to \infty$. The equation of a general Hoerl function is as follows:

$$f(x) = ab^x x^c$$

The parameter a is responsible for the vertical stretch of the function. I did not find the roles of parameters b and c regarding the shape of a Hoerl function. From a modelling point of view, the initial increase in serum concentration is accounted for by the parameter a and by the power function term x^c. The subsequent decrease and plateauing in serum concentration is accounted for by the decaying exponential term b^x, where $0 < b < 1$. Fundamentally, this particular nonlinear regression model was chosen as it provides an accurate representation of the shape of a pharmacokinetic profile.

[8] Hyams, D., 2020. *Hyams Development*. [online] Curveexpert.net. Available at: <https://www.curveexpert.net/> [Accessed 25 September 2020].

To six decimal places, Curve Expert yielded the following values for parameters a, b and c:

$$a \approx 301.161357$$

$$b \approx 0.563576$$

$$c \approx 1.570177$$

I then substituted the values of these parameters into the equation of a general Hoerl function. Therefore, according to this Hoerl model, the equation of my pharmacokinetic profile for erythromycin is given as:

$$C(t) = 301.161357 \times 0.563576^t t^{1.570177}$$

This function can be graphed on SciDAVis to show how accurately it models the data I obtained for the pharmacokinetic profile of erythromycin.

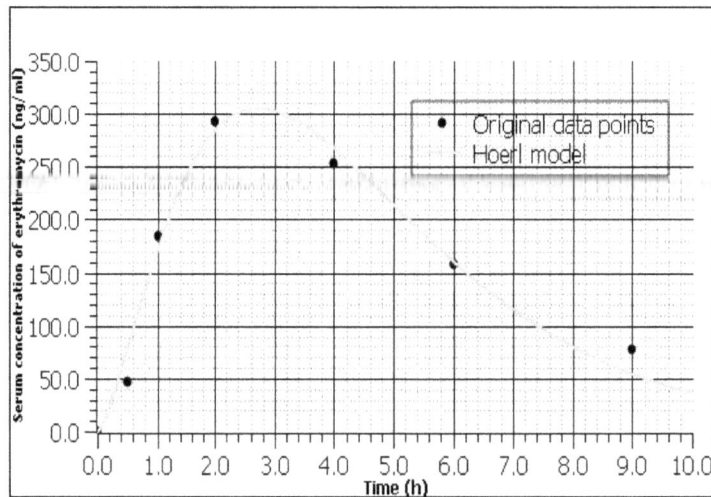

Fig. 5. Hoerl model (green) compared to the original data points (black).

To compare the C_{max} of my Hoerl model to that of the data, I will use the same procedure I used for my cubic model. However, this time I will use Wolfram Alpha's[9]

[9] Wolframalpha.com. 2020. *Derivative Of 301.161*0.564^(X)X^(1.570) - Wolfram|Alpha*. [online] Available at: < https://www.wolframalpha.com/input/?i=derivative+of+301.161357*0.563576%5E%28x%29x%5E%281.570177%29 > [Accessed 25 September 2020].

154

online calculator to find the first derivative of my Hoerl model, rather than calculate it

manually, for conciseness.

$$\frac{d}{dt}(301.161357 \times 0.563576^t t^{1.570177}) = 0.563576^t(472.877t^{0.570177} - 172.702t^{1.57018})$$

At the C_{max}, $0.563576^t(472.877t^{0.570177} - 172.702t^{1.57018}) = 0$

Using my GDC to sketch the function and find its roots,

$$t \approx 2.738101$$

Thus, the C_{max} of my Hoerl model is given by $C(2.738101)$, which ≈ 306.7 ng/ml.

Needless to say, a disadvantage of my Hoerl model with regards to my cubic model is

that it yields a C_{max} which is relatively greater than the C_{max} of the data. This is relevant

because if erythromycin's concentration surpasses a certain value (the MTC), it

carries the most prominent risk of cardiotoxicity among the more commonly used

macrolide antibiotics (Farzam et al.).[10]

I will now find the AUC of my Hoerl model using definite integration and compare this

to the AUC of my cubic model.

$$AUC = \int_0^9 (301.161357 \times 0.563576^t t^{1.570177})\, dt$$

A GDC was used to find the value of this definite integral, giving:

$$AUC \approx 1632 \text{ h ng ml}^{-1} \text{ (rounded to the nearest whole integer)}$$

This can be shown graphically using SciDAVis.

[10] Farzam, K., Nessel, T. and Quick, J., 2020. *Erythromycin*. [online] Ncbi.nlm.nih.gov. Available at:
<https://www.ncbi.nlm.nih.gov/books/NBK532249/#:~:text=Erythromycin%20carries%20the%20most%20prominent,is%20reco
mmended%20to%20minimize%20risk.> [Accessed 15 November 2020].

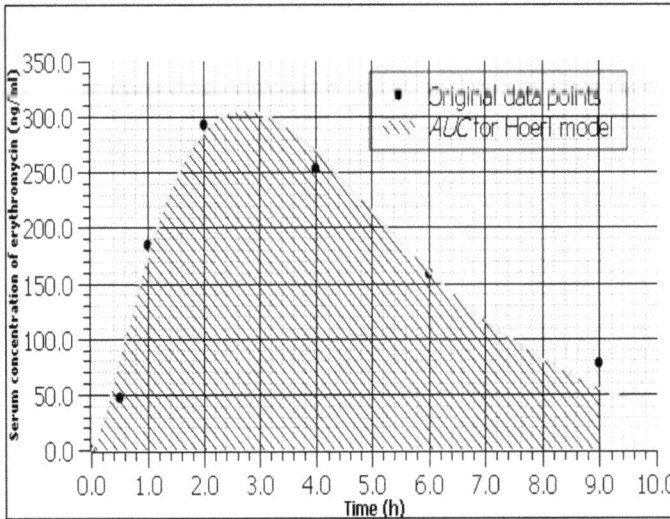

Fig. 6. Graphical representation of the AUC for my Hoerl model.

While the C_{max} of my two models are relatively similar, the AUC values are not. This is relevant because some antibiotics are dosed using AUC to quantitate the maximum tolerated concentration (AUC dosing). The efficacy of some antibiotics is related to AUC/MIC (minimum inhibitory concentration), thus maintaining a concentration above MIC is more important than peak concentrations (Bourne).[11] MIC is defined as the lowest concentration of an antimicrobial that will inhibit the visible growth of a microorganism after overnight Incubation (Andrews).[12] On balance, I consider my Hoerl model to be a more valid fit of the data overall than my cubic model. My Hoerl model represents the behaviour of the serum concentration of antibiotic over time more accurately overall, particularly because it has the time axis as a horizontal asymptote. In other words, my Hoerl model renders an overall better pharmacokinetic profile of oral erythromycin than my cubic model. To further evaluate the reliability of my cubic model and Hoerl model, I will compare the observed serum concentration of erythromycin to the cubic model and Hoerl model predicted concentrations of erythromycin. As the maximum number of decimal places in the observed concentrations is one, the model predicted

[11] Bourne, D., 2020. *Area Under the Plasma Concentration Time Curve (AUC)*. [online] Boomer.org. Available at: <https://www.boomer.org/c/p4/c02/c0208.php#:~:text=Toxicology%20AUC%20can%20be%20used,as%20shorter%20but%20higher%20concentration.> [Accessed 26 September 2020].
[12] Andrews, J., n.d. *Determination Of Minimum Inhibitory Concentrations*. [online] https://pubmed.ncbi.nlm.nih.gov/. Available at: <https://pubmed.ncbi.nlm.nih.gov/11420333/> [Accessed 1 November 2020].

concentrations will also be given correct to one decimal place. Moreover, to accentuate how my Hoerl model is more suitable than my cubic model, I will calculate the percentage error of each model, using the formula shown below.

$$\% \text{ error} = \left| \frac{\text{Observed concentration} - \text{Model predicted concentration}}{\text{Observed concentration}} \right| \times 100$$

Time (h)	Serum concentration of erythromycin (ng/ml)	Cubic model	% error	Hoerl model	% error
0.0	0.0	0.0	0.0	0.0	0.0
0.5	47.4	105.9	123.3	76.1	60.5
1.0	185.0	185.0	0.0	169.7	8.3
2.0	293.0	274.2	6.4	284.0	3.1
4.0	254.0	254.0	0.0	267.9	5.5
6.0	158.0	116.4	26.3	160.8	1.8
9.0	76.9	77.2	0.0	54.4	29.3

Table 5- Observed and model predicted serum concentrations of erythromycin, including the percentage error for each model.

As stated earlier, although the cubic model is a good fit for the data up to $t = 7.5$, it then becomes completely invalid after this time. This is because for very large values of t, $P(t)$ approaches ∞. Also, the cubic model predicted serum concentration increases too rapidly at first, as shown by the high % error of 123.3% at $t = 0.5$.

Notwithstanding the fact that my Hoerl model is a suitable model for the data, one of its limitations is that from 0 to 0.5 hours, the Hoerl model predicted concentration of erythromycin increases too rapidly, as shown by the high % error of 60.5. Also, from 6 to 9 hours, the Hoerl model predicted concentration of erythromycin decreases too rapidly. Nevertheless, I am reasonably satisfied with using my Hoerl model to depict the pharmacokinetic profile of erythromycin from the study mentioned earlier. I used SciDAVis to graph my Hoerl regression model superimposed over the graph of the original data points and of my cubic model.

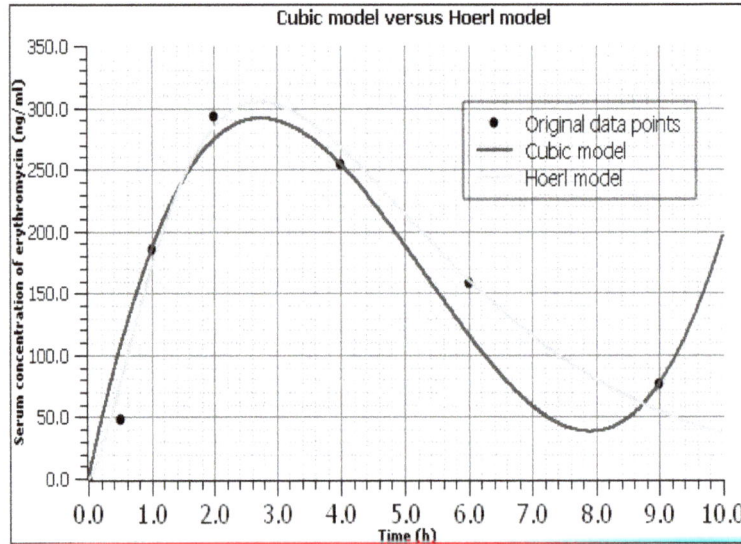

Fig. 7. Comparison of my Cubic model (blue) versus my Hoerl model (green) to represent the pharmacokinetic profile of erythromycin (the original data points are shown in black).

Conclusion

Conclusively, even though I was able to fulfil my aim, there are clear limitations with using polynomial functions to model a pharmacokinetic profile. Firstly, despite the fact that my cubic model has a similar C_{max} to the data, it suggests that as $t \to \infty$, the serum concentration of erythromycin also $\to \infty$. As shown in *Figure 1*, the serum concentration plateaus as $t \to \infty$. My cubic model does not account for this. Consequently, when I cross-checked my cubic model with basic pharmacokinetic principles, it was immediately apparent that it was not a valid model for the data. Accordingly, and even though my Hoerl model accounts for the plateauing in the serum concentration of erythromycin as $t \to \infty$, an extension of this exploration could be to come up with a model which illustrates this pharmacokinetic principle, using distribution and elimination pharmacokinetic parameters which are specific to each drug. A more suitable function to model this data would be a biexponential decay function in the form of

$$C(t) = Ae^{-\alpha t} + Be^{-\beta t}$$ proposed by Nigrovic[13],

Where:

$C(t)$: Serum concentration at any time after injection.

A and B: extrapolated (to $t = 0$ hours) concentrations in the serum.

α and β: first-order rate constants associated with the distribution and elimination phases respectively, of the serum concentration-time curve (Baggot) [14], where $\alpha > \beta$ by definition.

A graphical representation of a biexponential decay function would look similar to the one suggested by Bourne[15]:

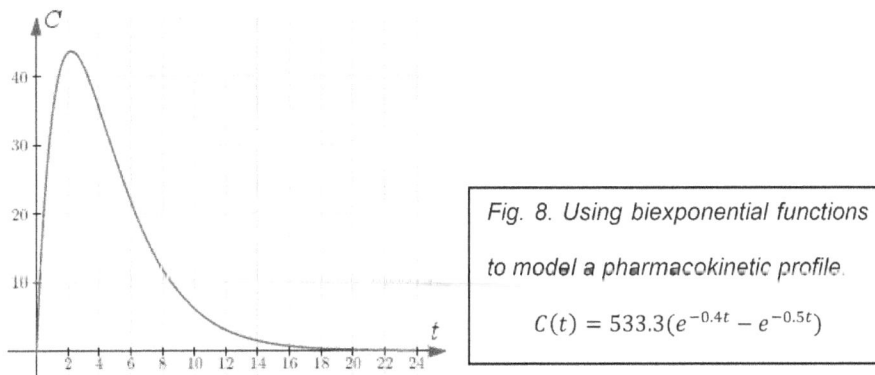

Fig. 8. Using biexponential functions to model a pharmacokinetic profile.

$$C(t) = 533.3(e^{-0.4t} - e^{-0.5t})$$

[13] Nigrovic, V., 1993. *PLASMA DRUG CONCENTRATIONS: DESCRIPTION AND INTERPRETATION OF THE BIEXPONENTIAL DECAY.* [online] Bjanaesthesia.org. Available at: <https://bjanaesthesia.org/article/S0007-0912(17)45613-3/pdf> [Accessed 27 August 2020].
[14] Baggot, J., 2001. *Pharmacokinetic Terms: Symbols And Units.* [online] https://onlinelibrary.wiley.com/. Available at: < https://onlinelibrary.wiley.com/doi/full/10.1046/j.1365-2885.2001.00340.x > [Accessed 26 September 2020].
[15] Bourne, M., 2010. *Math Of Drugs And Bodies (Pharmacokinetics).* [online] Intmath.com. Available at: <https://www.intmath.com/blog/mathematics/math-of-drugs-and-bodies-pharmacokinetics-4098> [Accessed 27 August 2020].

Personally, an interesting extension to this investigation would be to graph the pharmacokinetic profile of erythromycin for different administration forms, such as intravenous and intramuscular. Then, I could come up with biexponential models for the different pharmacokinetic profiles and find the AUC for each one using integral calculus. This would enable me to compare how the AUC values for erythromycin differ with different administration forms. If my family member had pneumonia again, I would advise him to use the administration route for erythromycin which offers the greatest bioavailability (i.e., the greatest AUC).

Furthermore, to satisfy my ardent curiosity for the topic of antibiotic resistance, it would be relevant to compare the pharmacokinetic profiles of erythromycin in individuals with high levels of erythromycin-resistant bacteria versus in individuals with low levels of erythromycin-resistant bacteria. This is a pertinent global issue because the addition of antibiotics to feed in factory farms constitutes a number of human health consequences, such as the emergence of resistant bacterial infections which cost healthcare systems vast amounts of money per year. As claimed by Michael Greger, M.D., "when animals receive unnecessary antibiotics, bacteria can become resistant to the drugs, then travel on meat to the store, and end up causing hard-to-treat illnesses in people."[16] This investigation would help shed light on whether erythromycin behaves differently following administration in the bloodstream of individuals with high levels of erythromycin-resistant bacteria versus in individuals with low levels of erythromycin-resistant bacteria.

---

[16] Greger, D., 2016. *Antibiotic-Resistant "Superbugs" In Meat | Nutritionfacts.Org*. [online] NutritionFacts.org. Available at: <https://nutritionfacts.org/2016/03/10/antibiotic-resistant-superbugs-in-meat/> [Accessed 17 August 2020].

Overall, it was fascinating to apply modelling (namely the Lagrange interpolation formula, which was an unfamiliar concept in mathematics to me), nonlinear regression, differential calculus, and integral calculus to pharmacokinetics, which is one of the major branches of pharmacology (the other one being pharmacodynamics). Pharmacology is one the subjects studied in medical school so by having done my exploration on this topic, I believe I will have a better understanding of the basic pharmacokinetic parameters when I hopefully come to study this in the future.

7. MODELING THE TIDE AND ITS CONNECTION WITH THE MOON PHASES

Author: Marilena Rentzou
Moderated Mark: 19/20
Level: Math AI SL

Introduction

In this project, I will try to create several mathematical models, using my knowledge of functions, their features and graphs, that will describe how water levels change over time during a day and also highlight the connection between tides and moon phases.

My reasons for doing so are twofold: Firstly, in order to attempt to construct a tide model that visitors to Phuket can consult before deciding to go for a swim or snorkeling and secondly to satisfy my curiosity and verify that "The disparity between high and low tide is at its highest during a full moon or a new moon."

More specifically, in this project, I will generate models describing how the water level rose and fell in Phuket, Thailand, on October 24th, 2020, from midnight to 23:00. I will then compare the model that provides the best fit (to the data of the 24th) against raw (observed) water level data for the same time period from October 25th in order to determine if the model provides as good a fit to this data set too and can, thus, be said to hold for a lengthier time period (to be determined) and not just a single day. I will also attempt to demonstrate the influence of the Moon on Earth's tides by looking at tide data on a new and full Moon.

The inspiration behind this internal assessment came from a trip I took to Phuket in Thailand two years ago. Here, for the first time in my life, I had to consider water levels before planning a beach activity. Due to the fact that the phenomenon of tides is not that noticeable in Greece where I live, except for the city of Chalkida, I was intrigued and decided to investigate how tides behave and to what extent do lunar cycles play a role in the process of their formation.

Tide is one of the most interesting natural phenomena in the world: it is the usual rise and fall of waters in the ocean. During high tide the water rises to a maximum level and reaches most of the shoreline. When water falls to its lowest level and recedes from the shore, that is low tide[1].

- [1]"Tide." *Dictionary.com*, Dictionary.com, www.dictionary.com/browse/tide. [Accessed 17 October 2020].

In general, within a twenty-four-hour cycle, most shorelines encounter two high and two low tides, while some places have only one each. The scale of the tides, high and low, varies during the month. The disparity between high and low tide is at its highest during a full moon or a new moon. That is because the gravitational pull is at its strongest when the Earth, the Sun and the Moon are in a line. The Sun's gravitational pull also plays a minor part in tide formation. Tides shift across the World as the ocean bulges.[2]

Fig.2: The phenomenon of tides in Phuket

But let us first look at what causes tides?

The tide phenomenon and what causes it

The gravitational field of the Moon is so strong it affects the Earth, particularly the ocean water. There is be a noticeable bulge on the Earth's side nearest to the Moon. The rise and

---

- [2]"Tide." *Wikipedia*, Wikimedia Foundation, 26 August. 2020, en.wikipedia.org/wiki/Tide. [Accessed 17 October 2020].

fall of the ocean level are a result of the pulling of the Moon's gravitational field as it moves around the Earth in orbit.

The tide is mainly created by the influence of the Moon's gravity, the sun's gravity, the rotation of the earth, the Moon's inclination, and meteorological conditions. The tide thus varies from place to place, due to the variance of the conditions mentioned above. The maximum value of the level rise is called high tide, and the minimum is called low tide.

While the Moon traverses its lunar orbit around the earth, at some point, it aligns with the Sun and the Earth twice, presenting the so-called conjugations and counterbalances.

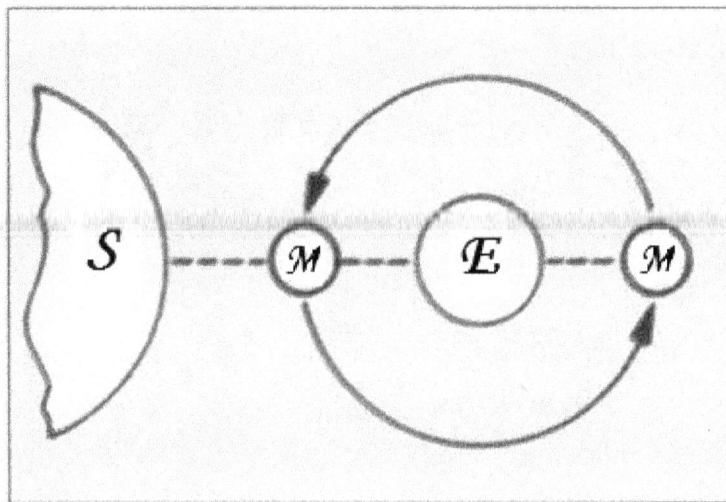

Fig.2 During its orbit around the Earth, the Moon aligns with the Earth and the Sun twice

When this happens, the gravitational forces of the sun and the moon are added, respectively increasing and decreasing the tidal effect. The high tides reach their maximum value and, respectively, the low tides to the minimum - Then we say that we have the spring tides (fig.1).

When the moon orbits the earth, it forms a right angle with the earth and the sun twice, that is, during its first and third quarters (fig. 2). When this happens, the gravitational forces of the sun and the moon are neutralized by each other, and therefore the intensity of the phenomenon is reduced. The high tides reach their minimum value and, respectively, the

166

low tides, the maximum. Then we can say that we have the square tides (neap tides). The tide range is the difference in tide height between high and low tide.

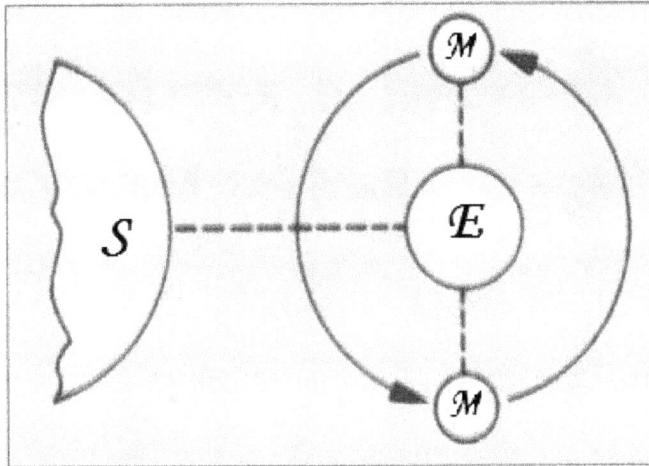

Fig. 3 During its orbit around the Earth, the Moon forms a right angle with the Earth and the Sun twice

The phases of the moon can be used to predict the tides of conjunction and those of squaring. When the Moon is situated between the Earth and the Sun either between the two (New Moon) or on the other side of the Earth (Full Moon) we know that we will have a tide of conjunction or opposition.

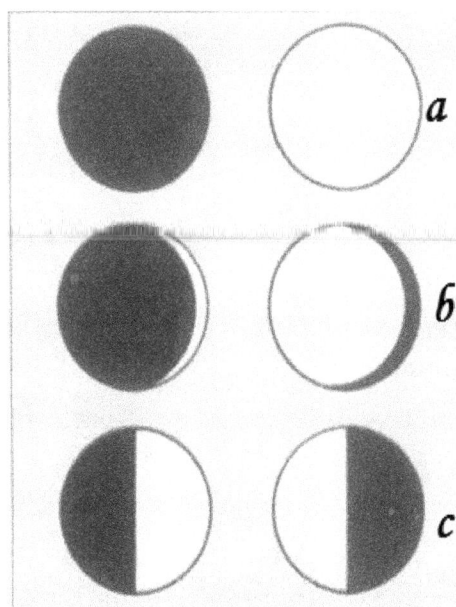

Fig. 4: The phases of the Moon

As we have previously mentioned, other factors that affect the tide are the gravity of the Sun, the local conditions, and even the rotation of the earth itself, which revolves around an axis that is not perpendicular but forms an angle to the Sun. These and other factors of lesser importance affect the behavior of the tide, which does not appear the same in all places. We have differences in duration between high tide and low tide, in range, etc. These differences identify three main types of tides, semi-diurnal, mixed, and diurnal. The half-day lasts about 6.25 hours, so there is a complete cycle from flood to shallow and flood again in half a day. The tidal range of this type rarely changes between successive cycles.

Tidal Charts and forecast

Worldwide, government agencies calculate tides on a daily basis to determine when the two highest and lowest tides will occur in a city and how high they will be. This data collection is called a "tide table" or "tide chart."

In general, people tend to want to predict, and this is due to their nature. The tides are one phenomenon that can normally be forecast fairly accurately. But getting it right demands hard work, and various factors are implicated. As previously mentioned, the tides are caused by the gravitational pull of the Sun and Moon, which create "bulges" of water in the oceans. So, the Sun and Moon positions — their distance from Earth, their direction in space, and how they move — are the most important factors in predicting the tides. But even more, is required for the most precise predictions. For starters, forecasters must know the contours of the seafloor. Climate conditions must also be identified to forecasters, as winds and air pressure will also influence the level of the water.

The raw data

Raw data for October 2020 was collected from the Royal Phuket Marina website (https://www.royalphuketmarina.com/marina/phuket-tide-tables/#10)

Table 1: Water Levels at Phuket during the month of October, 2020.

To construct a first model describing how the water level rises and falls during a day the data for October 24th will be used.

Hours after midnight	Height of water in meters 24-10-2020	Height of water in meters 25-10-2020	Height of water in meters 17-10-2020 -New moon	Height of water in meters 31-10-202 -Full moon
0	2.2	2.0	3.1	2.8
1	2.4	2.0	2.6	2.3
2	2.5	2.2	1.9	1.7
3	2.6	2.3	1.2	1.2

169

4	2.6	2.4	0.8	1.0
5	2.5	2.5	0.7	1.1
6	2.3	2.5	1.0	1.4
7	2.1	2.4	1.6	1.9
8	1.9	2.3	2.4	2.4
9	1.8	2.1	3.0	2.9
10	1.7	1.9	3.4	3.1
11	1.7	1.8	3.5	3.0
12	1.8	1.7	3.2	2.7
13	1.9	1.7	2.7	2.3
14	2.0	1.7	2.0	1.7
15	2.1	1.9	1.2	1.3
16	2.2	2	0.7	1.0
17	2.3	2.2	0.5	1.0
18	2.3	2.3	0.8	1.3
19	2.2	2.4	1.4	1.9
20	2.1	2.4	2.1	2.5
21	2.0	2.3	2.9	3.0
22	2.0	2.2	3.4	3.3
23	2.0	2.0	3.6	3.3

Table 2: Raw data used in this investigation

Mathematical Analysis

To decide on a suitable model height, $\underset{\text{dependent variable}}{\underline{h \ (in \ metres),}}$ will be plotted against time

$\underset{\text{independent variable}}{\underline{t \ (in \ hours),}}$ since midnight:

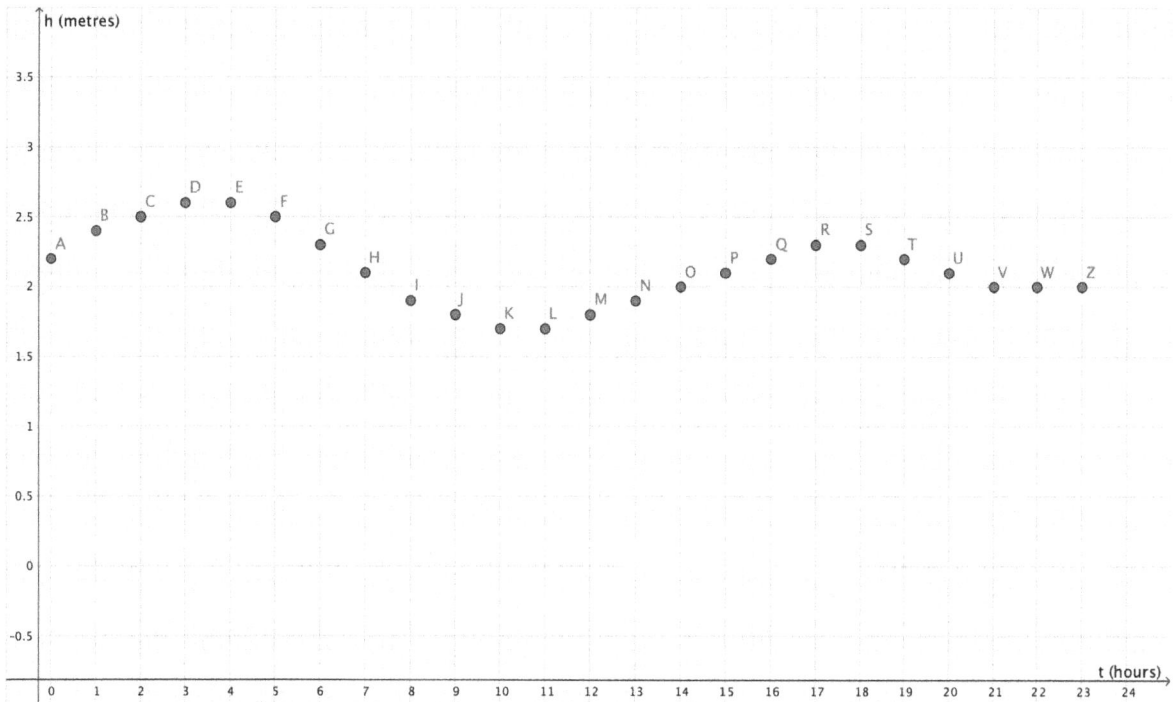

Graph 1: Raw data for October 24ᵗʰ, 2020

Looking at the shape «drawn» by the raw data points I decided to develop a sinusoidal model to describe the tide phenomenon on October 24th, 2020 in Phuket:

The general form of the sinusoidal function is $h(t) = a \cdot \sin(b(t + c)) + d$, where $a \in \mathbb{R}$, $b \in \mathbb{R}, \ c \in \mathbb{R} \ and \ d \in \mathbb{R}$ are parameters that will be found using the raw data.

$|a|$ affects the amplitude (height) of the curve

b affects the number of full sine curves viewed between 0 and 2π (or $360°$). The period T of the sine function can then be found using the formula: $T = \dfrac{2\pi}{h}$

c affects the horizontal shift of the curve from the y-axis

d affects the vertical shift of the curve from the x-axis

As a first step I will find a and d:

I know from theory that

$-1 \leq \sin(b(t + c)) \leq 1$

I also know that I can multiply an inequality across by a positive constant $|a|$ without affecting the direction of the inequality

171

Hence,

$$-|a| \leq |a| \cdot \sin(b(t+c)) \leq |a|$$

I can also add the same constant quantity d to all terms of the inequality without affecting the direction of the inequality.

I can, thus, write

$$-|a| + d \leq |a| \cdot \sin(b(t+c)) + d \leq |a| + d$$

In the inequality above the quantity $|a| + d$ represents the maximum value of the function $h(t) = a\sin(bt+c) + d$,

while

the quantity $-|a| + d$ represents the minimum value of the function $h(t) = a\sin(bt+c) + d$,

Looking at the raw data table, the least recorded water height value for October 24th 2020 is 1.7 m and the greatest recorded value on the same day is 2.6:

I will use these two values to construct two linear equations in two unknowns in order to find the values of a and d:

$$|a| + d = 2.6 \; [Equation \; 1]$$
$$-|a| + d = 1.7 \; [Equation \; 2]$$

Adding the two equations I obtain: $2d = 4.3 \Rightarrow d = 2.15$

Substituting $d = 2.15$ into $Equation$ 1 and solving will allow me to find the value of $|a|$:

$$|a| + 2.15 = 2.6 \Rightarrow |a| = 0.45$$

Since the curve first rises and then falls I will let a be positive.

So now: $h(t) = 0.45 \cdot \sin(b(t+c)) + 2.15$. Next, I will find b:

Looking at the raw data graph, I observe that the time between a maximum point (4, 2.6) and the consecutive minimum point (10, 1.7) is 6 hours. This corresponds to half a period for the tide phenomenon.

That is, period $T = 2 \cdot 6 = 12$. But, $: T = \frac{2\pi}{b}$.

172

Therefore: $\underbrace{\dfrac{2\pi}{b} = 12}_{Equation\ 3} \Longrightarrow b = \dfrac{2\pi}{12} = \dfrac{\pi}{6}$.

I now have that $h(t) = 0.45\sin\left(\dfrac{\pi}{6}(t+c)\right) + 2.15$

To find c, I will choose a point from my raw data to plug into the equation for $h(t)$:

I chose the point (0, 2.2).

Substituting into $h(t)$ I get:

$$2.2 = 0.45 \cdot \sin\left(\dfrac{\pi}{6}(0+c)\right) + 2.15 \ [Equation\ 4]$$

I will now solve this equation to find c:

$$2.2 = 0.45 \cdot \sin\left(\dfrac{\pi}{6}(0+c)\right) + 2.15 \Longrightarrow 2.2 - 2.15 = 0.45 \cdot \sin\left(\dfrac{\pi}{6}\cdot c\right)$$

$$\Longrightarrow 0.05 = 0.45 \cdot \sin\left(\dfrac{\pi}{6}\cdot c\right) \Longrightarrow \dfrac{1}{9} = \sin\left(\dfrac{\pi}{6}\cdot c\right)$$

$$\Longrightarrow \sin^{-1}\left(\dfrac{1}{9}\right) = \dfrac{\pi}{6}\cdot c \Longrightarrow \dfrac{6}{\pi} \cdot \sin^{-1}\left(\dfrac{1}{9}\right) = c$$

$$\Longrightarrow c = \dfrac{6}{\pi} \cdot \sin^{-1}\left(\dfrac{1}{9}\right) \Longrightarrow c = 0.213(3s.f)$$

Thus, $\boldsymbol{h(t) = 0.45\sin\left(\dfrac{\pi}{6}(t+0.213)\right) + 2.15,\ 0 \le t \le 23}$

In order to appreciate how well this model fits it[3], I will plot

$h(t) = 0.45\sin\left(\dfrac{\pi}{6}(t+0.213)\right) + 2.15,\ 0 \le t \le 23$ against the raw data:

---

[3] The raw data

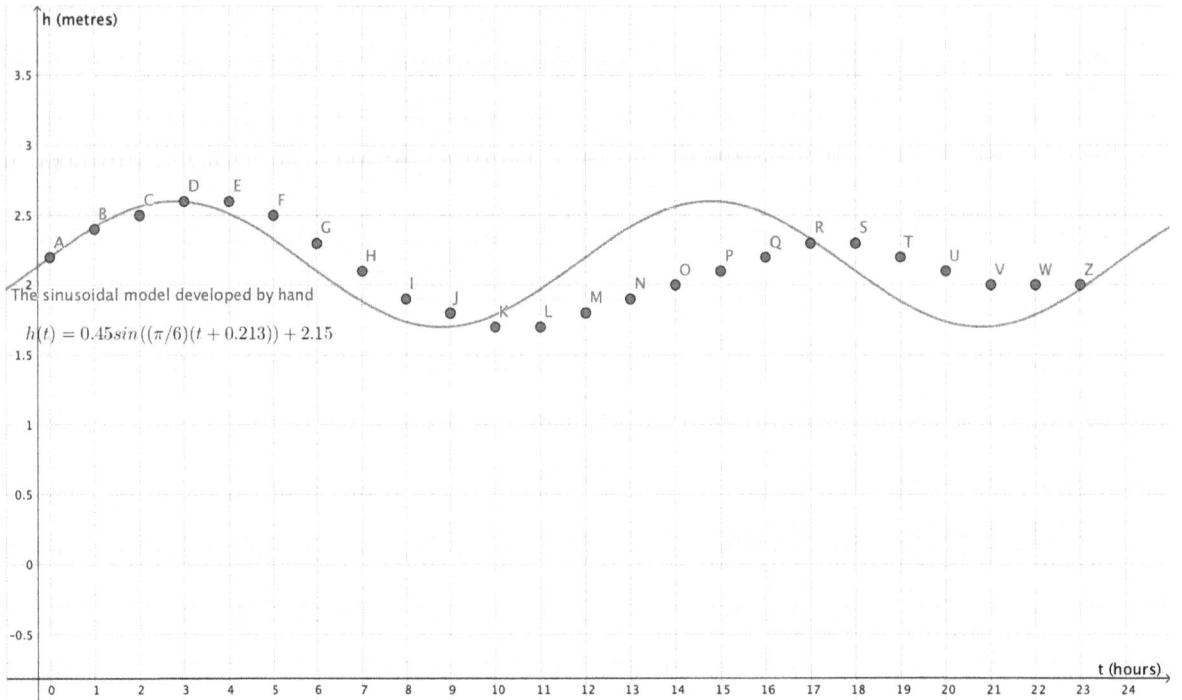

Graph 2: Manual sinusoidal model

Though the model passes through the first few points and describes the general trend, it does not provide a good fit for the raw data after 3.00 a.m.

To provide a better model I will employ the aid of technology and, specifically, Geogebra's FitSin(<List of Points>) feature:

The resulting model can be seen below:

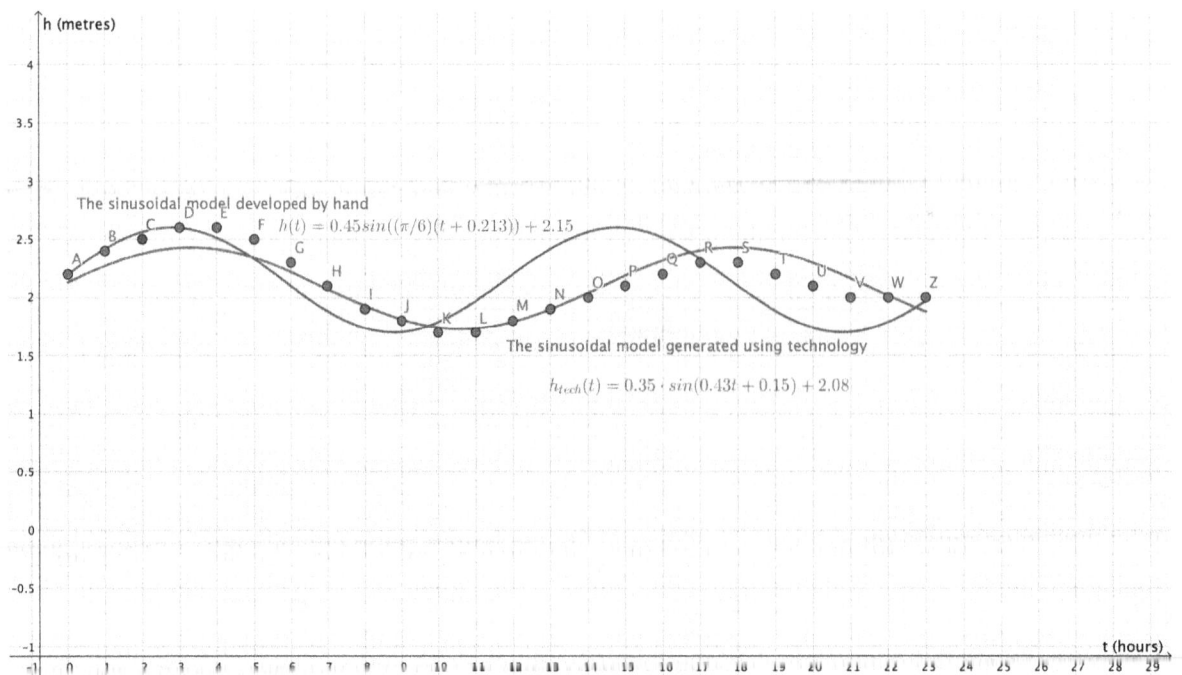

Graph 3: Manual sinusoidal model vs sinusoidal model derived using technology

The sinusoidal model, $h(t)_{tech} = 0.35 \cdot \sin(0.43 \cdot t + 0.15) + 2.08, \ 0 \leq t \leq 23$,

generated using technology obviously fits the data better, especially after 6.00 a.m.

To further check this models performance I will plot it against the raw data for October 25th,

2020:

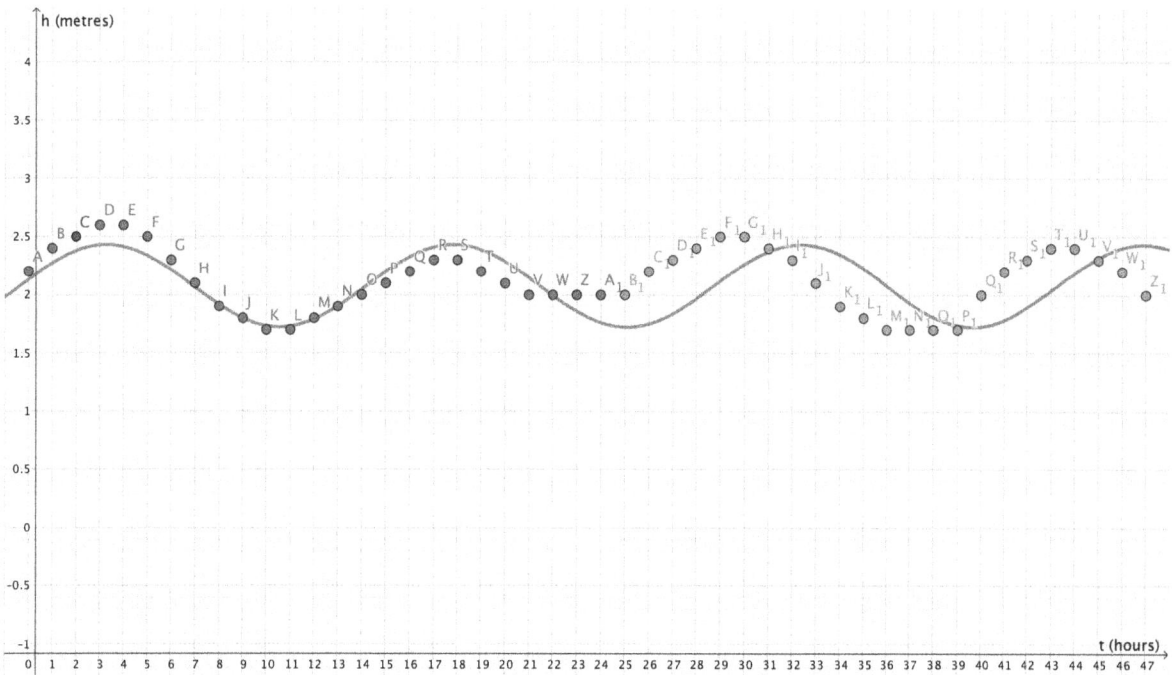

Graph 4 : Raw data for October 24th, and October 25th, 2020

In the graph above the blue data represent the water levels on October 24th, 2020 from

midnight to 23:00 and the lilac points represent the water levels on October 25th, 2020 from

midnight to 23:00

Unfortunately, $h(t)_{tech} = 0.35 \cdot \sin(0.43 \cdot t + 0.15) + 2.08, \ 0 \leq t \leq 23$ does not fit the

data for October 25th, 2020 well.

This is only to be expected as "The difference in sea-level height between each high and low tide changes daily depending upon the position of the Moon."[4]

Using Geogebra's FitSin(<List of Points>) feature the model
$h(t)_{tech2} = 0.34 \cdot \sin(0.47 \cdot t - 0.07) + 2.1, \ 0 \leq t \leq 47$ is generated which fits the raw data over the two days well as can be seen below – But I suspect as a result of my previous findings and research that this model will not fit the raw data for October 26th well.

"High and low tides, and their timing, for every location world-wide are different day by day, month by month and year by year."[5]

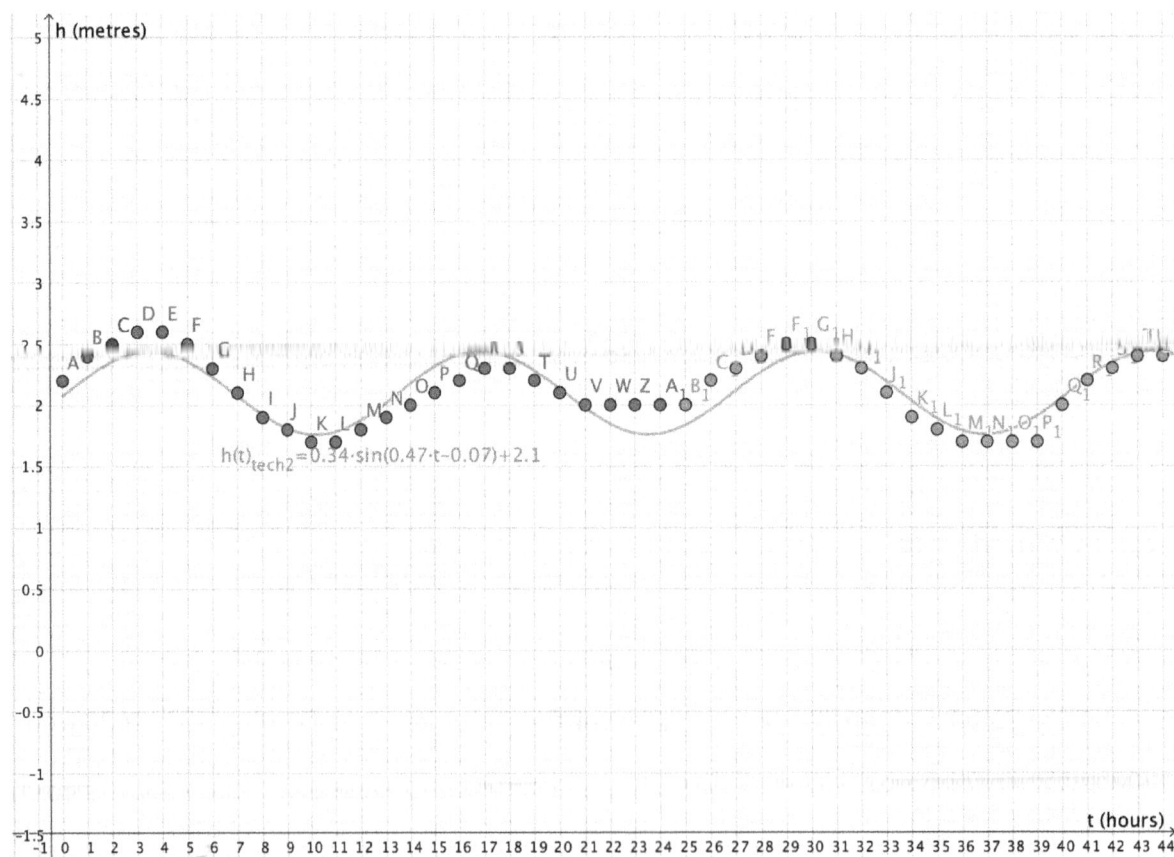

Graph 5 : Sinusoidal model derived using technology

---

[4] US Department of Commerce, NOAA. "Tides." *NWS JetStream*, NOAA's National Weather Service, 9 Aug. 2019, www.weather.gov/jetstream/tides. [Accessed 17 October 2020].

[5] US Department of Commerce, NOAA. "Tides." *NWS JetStream*, NOAA's National Weather Service, 9 Aug. 2019, www.weather.gov/jetstream/tides. [Accessed 17 October 2020].

To investigate whether indeed , "the disparity between high and low tide is at its highest during a full moon or a new moon", I plotted the water level data for October 17th, 2020 (New Moon) and for October 31st (Full Moon) against time since midnight. See Graph below:

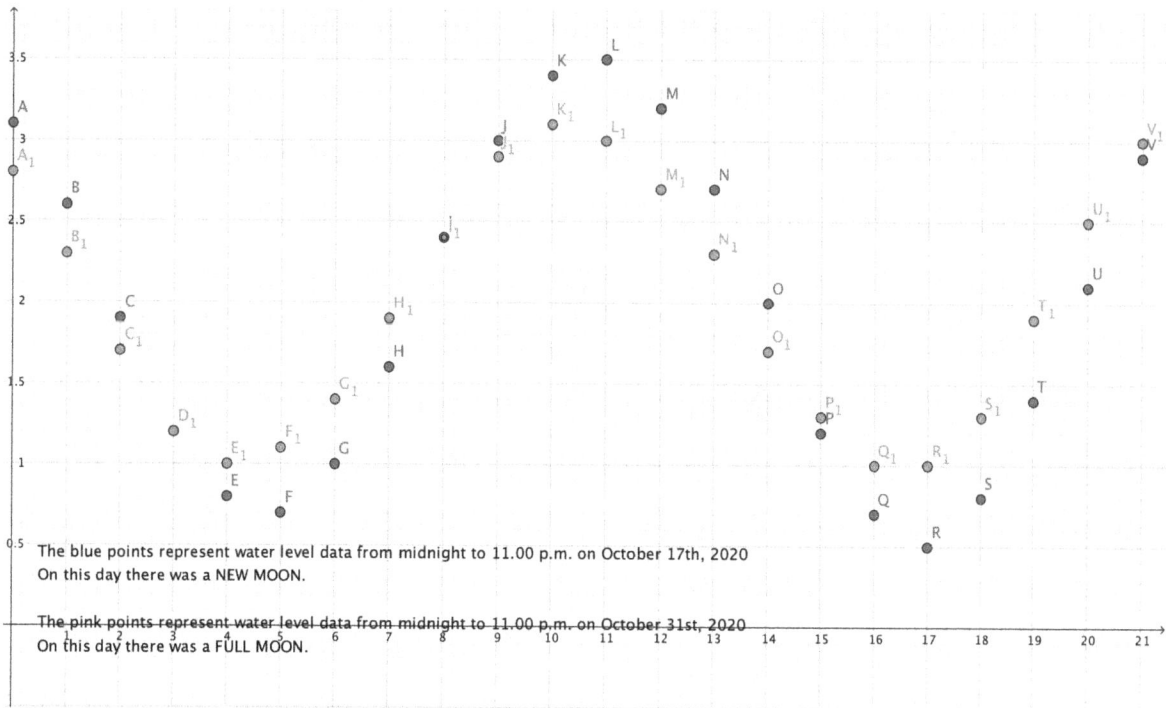

The blue points represent water level data from midnight to 11.00 p.m. on October 17th, 2020
On this day there was a NEW MOON.

The pink points represent water level data from midnight to 11.00 p.m. on October 31st, 2020
On this day there was a FULL MOON.

Graph 6: New Moon and Full Moon Raw Data

On October 17th the first low tide occurs at 5.00 a.m. At this time the water reaches a height of 0.7 m. and The first high tide occurs at 11.00 a.m. at which time the water reaches a height of 3.5 m. The difference in water levels is 2.8 m.

On October 31st the first low tide occurs at 4.00 a.m. At this time the water reaches a height of 1.0 m. and The first high tide occurs at 10.00 a.m. at which time the water reaches a height of 3.1 m. The difference in water levels is 2.1 m.

On October 24th, 2020 the difference in water levels between the first low and first high tides is just 0.9 m.

-Interestingly, high and low tides on these two days (17-10-2020, 31-10-2020) occur at approximately the same times.

Summary of results

Despite the fact that weather conditions and other factors that affect tides, were not taken into consideration I managed to create a model for the 24th of October 2020 as it can be seen in the exploration:

$$h(t)_{tech} = 0.35 \cdot \sin(0.43 \cdot t + 0.15) + 2.08, \quad 0 \leq t \leq 23$$

As the model above did not prove to be a good fit for the water level data for October 25th, a second model

$$h(t)_{tech2} = 0.34 \cdot \sin(0.47 \cdot t - 0.07) + 2.1, \quad 0 \leq t \leq 47$$

was also generated to fit the raw data over the 24th and 25th of October.

In *Graph 7* below the two models are plotted against the raw data for comparison

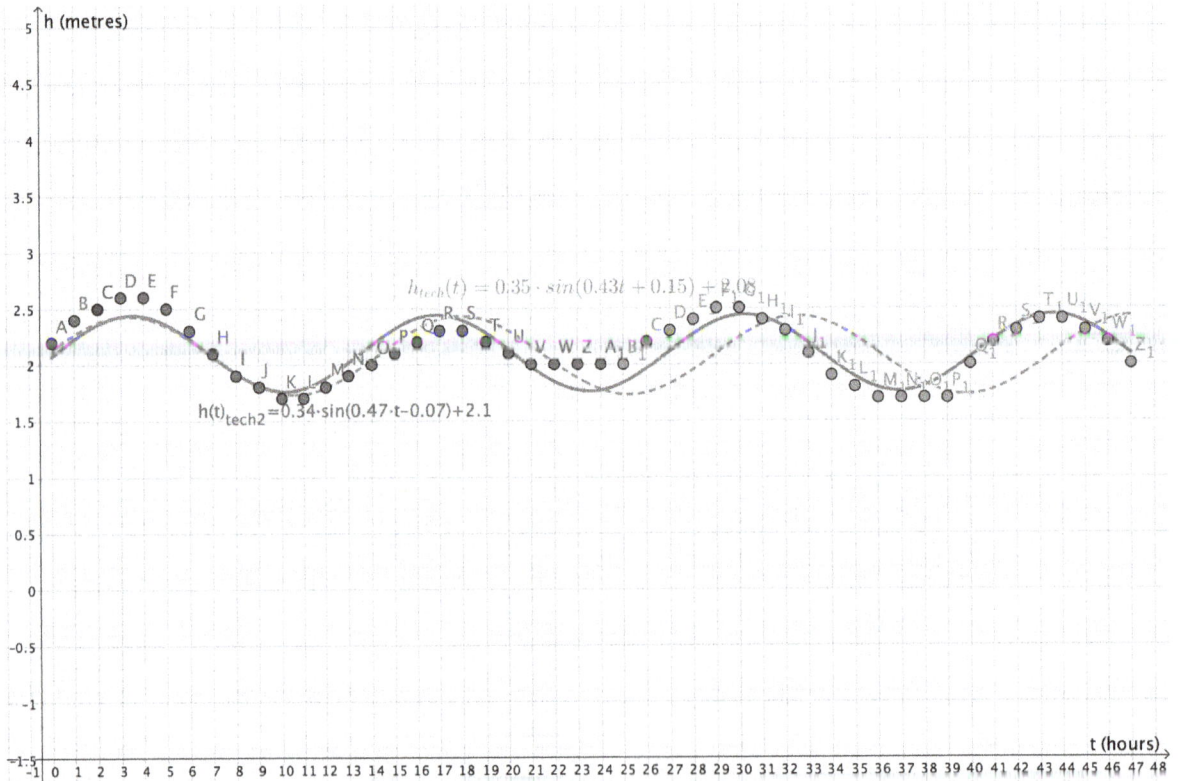

Graph 7: A comparison of models

Both models are closely aligned until 10.00 a.m. of October 24th, 2020. Both slightly under-predict the water level values between midnight and 7.00 a.m. After 10.00 a.m. the dotted orange line $(h(t)_{tech} = 0.35 \cdot \sin(0.43 \cdot t + 0.15) + 2.08, \ 0 \leq t \leq 23)$ continues to pass through the raw data points until 14.00. However, after 14.00 the predicted water level values begin, increasingly, to lag compared to the raw data. On the other hand, the full red line

$(h(t)_{tech2} = 0.34 \cdot \sin(0.47 \cdot t - 0.07) + 2.1,\ 0 \le t \le 47)$ follows the data more closely, though not entirely accurately -other times over-predicting (10.00 - 19.00 on 24-10-2020, 9.00 - 16.00 on 25-10-2020) and other times under-predicting (22.00 – 24.00 on 24-10-2020, 1.00 – 6.00 on 25-10-2020) actual water level values. It accurately, however, indicates the times water levels are at their highest/lowest points. But even this model only serves to describe water levels between October 24th, midnight and October 25th, 23.00.

Finally, raw data was used to demonstrate that, as expected from research, during new and full moons the difference between low and high tide water levels was greater than on other days. This observation provides evidence that the "Moon plays the biggest role"[6], influencing tides more than other factors.

Limitations

Despite the above-mentioned models that I found through my investigation it has to be mentioned that several limitations exist: There are several factors that have been found to be fundamental since they affect the level of the tides. Firstly, tides may be influenced by local winds and weather patterns as well. Powerful offshore winds, which exaggerate low tide visibility, may drive water away from coastlines. Additionally, the force of tides can also be magnified by the form of bays and estuaries. In particular, funnel-shaped bays can change the tidal magnitude drastically, and one example that demonstrates such an effect is The Bay of Fundy in Nova Scotia. Lastly, they will be affected, not only on marine environments but on coastal areas that are home to millions of people and animals, would be impacted by the increase in sea levels.

Discussion

---

[6]Vigdis Hocken, et al. "What Causes Ocean Tides?" *Timeanddate.com*, www.timeanddate.com/astronomy/moon/tides.html#:~:text=The%20greatest%20difference%20between%20 high,water%20in%20the%20same%20direction. [Accessed 6 January 2020].

Through this investigation, I realized that it is not possible to construct an accurate model that describes how water levels change over time during a day. This happens because weather conditions change on a daily basis and thus predictions can be neither concrete nor general. However, a more precise model that would contain factors affecting tides could be created but not by me since my mathematical skills as a Mathematics SL student are limited. Nonetheless, if I had more time, I would like to investigate more extensively the connection to moon phases.

References

I.   *Seasons, Tides, and Lunar Phases* Tara Haelle page 8-12 [Accessed 16 July 2020].

II.   *Pacific Shore Fishing,* Michael Sakamoto page 190-195 [Accessed 17July 2020].

III. US Department of Commerce, National Oceanic and Atmospheric Administration. *The Importance of Monitoring the Tides and Their Currents - Tides and Water Levels: NOAA's National Ocean Service Education*, 1 June 2013. [Accessed 17 July 2020].

IV.   oceanservice.noaa.gov/education/tutorial_tides/tides09_monitor.html. [Accessed 17 October 2020].

V.   "Learning Centre." *Tourism NB – Hopewell Rocks | Tide Tables*, www.thehopewellrocks.ca/index.php/en/page/tide-tables. [Accessed 19 October 2020].

VI.   National Geographic Society, Morgan Stanley. "Tide." *National Geographic Society*, 30 Aug. 2019, www.nationalgeographic.org/encyclopedia/tide/. [Accessed 19 October 2020].

VII. "What Are Tides?" *BBC Bitesize*, BBC, 5 Nov. 2020. [Accessed 19 October 2020]. www.bbc.co.uk/bitesize/topics/z8c9q6f/articles/zdqr97h. [Accessed 19 October 2020].

VIII. "Tides." *Met Office*, www.metoffice.gov.uk/weather/learn-about/weather/oceans/tides. [Accessed 19 October 2020].

IX.   Horton, Jennifer. "What Are Tide Tables?" *HowStuffWorks Science*, HowStuffWorks, 27 Jan. 2020, science.howstuffworks.com/environmental/earth/oceanography/tide-table.htm. [Accessed 19 October 2020].

X.   Benningfield, Damond. "Predicting Tides." *Predicting Tides | Science and the Sea*, 5 Aug. 2018, www.scienceandthesea.org/program/predicting-tides. [Accessed 17 October 2020].

XI.   "Trigonometric Functions | Algebra (All Content) | Math." *Khan Academy*, Khan Academy, www.khanacademy.org/math/algebra-home/alg-trig-functions. [Accessed 17 October 2020].

XII. "Unit Circle." *From Wolfram MathWorld*, mathworld.wolfram.com/UnitCircle.html. [Accessed 17 October 2020].

XIII. "Introduction to TidesThis Unit Discusses the Origin of Tides Due to the Gravitational Attraction of the Moon and Sun, Tidal Variations, and Monitoring Tides." *Tides*, teachearthscience.org/tides.html. [Accessed 17 October 2020].

XIV. "Education - NOAA Tides & Currents." *Tides & Currents*, tidesandcurrents.noaa.gov/education.html. [Accessed 17 October 2020].

XV. admin, Posted By: "5 Best Editable Mathematics Assignment Cover Pages: MS Word Cover Page Templates." *MS Word Cover Page Templates | Download, Personalize & Print*, 20 Nov. 2018, www.mswordcoverpages.com/mathematics-assignment-cover-pages/. [Accessed 4 March 2021].

www.ingramcontent.com/pod-product-compliance
Lightning Source LLC
Chambersburg PA
CBHW061104210326
41597CB00021B/3978